D0891180

(Maine Department of Marine Resources)

The World of the SMALL COMMERCIAL FISHERMEN

Their Lives and Their Boats

Michael Meltzer

Dover Publications, Inc.
New York

To Joel and to Ellen

The author is especially grateful to
Erica Meltzer, Spencer Severson
and Martin Conan Halpern
for supplying important information.

Published in Canada by General Publishing Company, Ltd., 30 Lesmill Road, Don Mills, Toronto, Ontario.
Published in the United Kingdom by Constable and Company, Ltd., 10 Orange Street, London WC2H 7EG.

The World of the Small Commercial Fishermen: Their Lives and Their Boats is a new work, first published by Dover Publications, Inc., in 1980.

International Standard Book Number: 0-486-23945-4
Library of Congress Catalog Card Number: 80-66783

Manufactured in the United States of America
Dover Publications, Inc.
180 Varick Street
New York, N.Y. 10014

Contents

Introduction

It is still dark as the crab boat backs out of its harbor slip. As it chugs past the breakwater, the eastern sky is turning a dim gray with the approaching dawn. The boat quivers gently from the throbbing diesel as it heads toward the open ocean. The icy wind works its way through windbreakers and sweaters.

Half an hour past the breakwater, the sun rises and the sky glows in red and gold. The ocean, calm at this hour, shines like an immense sapphire. It is exhilarating to be awake so early, to see the new day's beginning. It is romantic—for the first hour. Then the boat reaches the fishing grounds and the work begins. A buoy is gaffed, the line that plunges from it down into the depths is slipped onto the hydraulic hauler and a crab trap is pulled up from the bottom. As it is hauled clear of the water, the heavy steel trap is swung onto the boat's rail and emptied. Then it is rebaited and thrown back and a new trap is pulled. On and on, hour after hour traps are hauled, emptied and thrown back. By midafternoon men are tired and tempers are short. The work continues until it is too dark to see the buoys. At that time the boat returns to port. But the damaged traps must be repaired and the boat washed down before anyone can go home.

The fishermen I interviewed rarely sat still as we talked. They felt compelled, driven to continue working, for there is always something that needs doing on a fishing boat. Sanding, painting, mending nets, splicing lines, straightening bent hooks, whatever. A fisherman's life is based on work and those that can't work and work hard do not make it. I found fishermen quite willing to accept men of different values and lifestyles, as long as they knew how to work. There is no place on a boat for the indolent and the lazy.

I asked fishermen all over the country why they fished, and I received many answers. The most common went something like: "It's the only way I know of you can make a fairly good living and not have to deal with people." Working at sea saves them the hassle of coping with customers or fellow office workers. Most all who can afford it own their own boats, making them answerable to no one else on earth except themselves. If they feel like laying off for a day or coming in early, no one can stop them.

Fishing is competitive and fishing is a gamble, and both these things serve to spur fishermen on to great efforts. It is quite an honor to be a harbor's highliner, while those that don't catch much usually won't talk about it if they can help it. The chance that tomorrow, one might blunder into the biggest school of fish ever discovered is also important for keeping a man going. And yet, no matter how good a season a fisherman is having, he never seems to catch enough. He could always do a bit better.

Most fishermen are men, although the number of women fishermen is steadily increasing. Most women work as crew for a male skipper, often their husband or boy friend, but a small number have their own boats. The women who do fish work just as hard as the men, for the sea is no kinder to one sex than to the other.

I see many similarities between fishermen and isolated farmers. Both groups make their living harvesting food and both spend their working hours isolated from the rest of the world. Both are fiercely independent and proud of it, and in both groups I have found a gentleness that I feel comes from working close to, and in harmony with, nature.

Pacific Salmon Fishing

A Salmon Puller's Story

I get to the harbor around 5:30 in the morning because Floyd is mad as a hornet if I'm any later. We warm up the engine, check the gurdies, make sure the lines are hooked up right, and are out on the ocean by sunup.

Where do we troll? That depends on where the action is. Say one of Floyd's friends had a good day yesterday by Shelly's Cove. Floyd might decide to try his luck there. As he pilots the boat to the grounds, I snap the leaders—these short lines with hooks or jigs on the end—onto the main fishing lines and unreel the gurdies so that the lines are dragging behind us in the water. Floyd likes to be all set up by the time we get to the grounds.

We leave the radios on all day, and listen to how the other boats are doing. If we're not into the fish but some other boat is, we get a fix on him with the RDF—the radio direction finder—and head on over.

When we get to the grounds Floyd tells me not to bother him and begins searching for the fish. "The color of the water tells you where the plankton are, boy," Floyd has told me. "Like as not there's bait fish nearby feeding on the plankton. An' if we're lucky, there's salmon feeding on the bait fish." Floyd also keeps his eye on the depth sounder, notes how choppy the water is and watches what the other boats in the area are doing.

Floyd usually guides the boat in big circles around the trolling grounds. When a fish takes a hook the bell at the end of the line rings and I turn on the proper gurdy and reel him in. If you reel them in too fast the line will snap and you'll lose seventy dollars worth of gear and maybe thirty dollars worth of fish.

When the fish breaks surface I gaff it in the gill, haul it aboard and club it between the eyes. It quivers for a few minutes and then dies. You have to catch salmon to know how beautiful they are. They have a brilliance to them while they're alive, like they've been coated with varnish. They lose that shine a few minutes after they die. I think it's because a mucus layer under the skin dries up.

I clean them after they're dead. I hold them by the gill, slit their bellies from anal pore to gills and scoop the guts out. Right underneath the spine, down where the belly and intestines are, is a little pocket full of a bloody sort of material that is very poisonous. It has to be broken open with a knife and cleaned out or else it will taint the fish. A friend of mine once fed his dog a fish where this hadn't been done and the dog died.

On a busy day I might clean fifty fish. When the action slows down, Floyd and I sometimes nurse a bottle of Jack Daniels and shoot the breeze. Floyd does most of the talking. When fishing is poor he complains about his gear and curses the fish for not jumping onto his hooks, but he rarely attacks his boat and never says anything nasty about the ocean. He has a kind of awe for the sea. He respects it like he would his father or mother. "This boat's just a hole in the ocean, boy," he often tells me. "And a small hole at that. And them super-freighters you see, they're small holes too. Man has never built a ship that can conquer the ocean, and he never will. The best he can do is learn to live with it."

We generally head home about seven. Sometimes the wind picks up late in the day and whips up the sea until it's pretty choppy. I always get nervous when this happens because it makes crossing the bar dangerous. Most harbors have sand bars near their mouths, built by riptides. The bars often are just under the surface, and they shift drastically from day to day. Once, a big swell picked us up as we were crossing the bar. As we slid down into the trough I looked over the side and saw sand just under the water's

(Joyce O'Neal)

surface. Floyd saw it too, and was he scared! He muttered something under his breath. A curse, a prayer, I don't know which. The next swell was bigger and the trough that followed was deeper. We dropped into it and I heard a rasping sound as our keel ground the bottom. I gripped the cap rail with all my might, closed my eyes and prayed. Then we shot clear and chugged safely toward the harbor. I remember sighing with relief.

When we reach the harbor, we unload at the fish buyer's and then clean up the boat for the next day. It's about ten by the time I get home. I usually make a quick dinner and go right to bed. We work every day of the week unless the wind is blowing too hard to go out, so I need all the rest that I can get.

Fishermen are the hardest-working people I've ever known, but they're very gentle too. I think it's because of all the death they have to deal with. I dig fishing because when I'm on the sea, I know who I am. Sometimes I hate it and can't wait until the season is over. But I come back, year after year.

Life Cycle of a Salmon

Life for a salmon begins in the gravel bed of some river, perhaps hundreds of miles upstream from the ocean. Its life starts when the bright red, pea-sized egg spawned by its mother is fertilized by the milt squirted onto it by its father.

For several weeks the egg does not change noticeably. Then, an eye becomes visible inside the semitransparent egg. A little later, a line appears that will eventually become the backbone. After ten weeks, the baby fish breaks through the egg. Attached to its underside is a sac, the remains of the egg's yolk. The fish continues to receive its nourishment from this sac for another month or so. During this time it lives in the gravel like a worm.

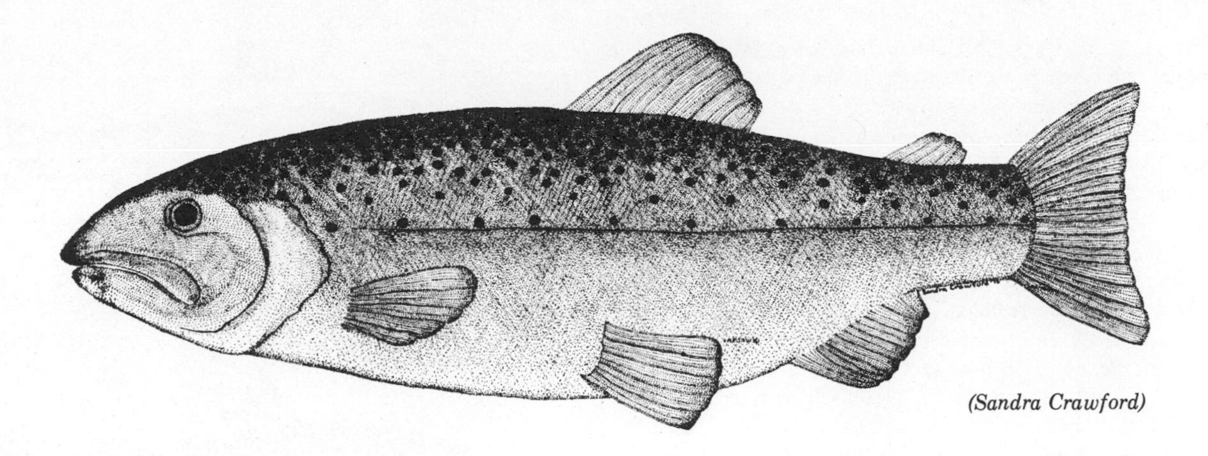

(Sandra Crawford)

When spring's warmth and sunlight come to the river, the little salmon come out of their burrows and feed on the numerous insects in the stream or get fed on by trout. The salmon resemble the trout at this point in their lives, for their backs are speckled with protective coloring.

Some wait only several months after they are born and some wait a full year, but sooner or later the salmon leave their place of birth and head toward the ocean. As the time of leaving approaches, changes occur. Their bodies soften. They stop eating and congregate in large groups. Suddenly the migration begins. They leap down falls and rapids in their dash to the sea. They swim through the turbines of hydroelectric plants unharmed, for they are still only a few inches in size. One of the biggest dangers they encounter at this time are irrigation ditches that lead from the river inland. The salmon sometimes follow them and get stranded. Birds also take many of the fish, especially after they have just leapt over a dam and lie stunned in the water below for a few seconds.

They pause when they get to the river's mouth, for their bodies must adjust to salt water. Eventually they enter the ocean with the outgoing tide. Little is known of their lives in the open sea, except that they travel great distances, growing fat on the ocean's abundant food supplies. When they are young they eat shrimp larvae and plankton, switching to herring and anchovy as they mature. They feed voraciously and can grow a foot and a half in a year.

After three or four years the adult fish, perhaps thirty pounds in weight, again begin to change. Their backs stiffen, their jaws develop a hook and bright coloration appears on their sides. They swim up and down the coast, trying to locate by sense of smell the river in which they were born. When they do find the river they swim up it, throwing themselves over rapids and leaping waterfalls as high as six feet. They don't eat during this time but live off their fat. By the time they reach the little stream or pool in which they were born, they are bruised and beat up and have lost most of their scales. They are no longer beautiful silvery fish, but are now black or dingy red.

When they reach the spawning areas the males fight and eventually pair off with females. They select spots for their nests and the females dig holes six to eighteen inches deep in the gravel bottom with their tails. Then each female lays a few hundred eggs in the holes and the male swims alongside her and squirts fertilizing milt onto them. After this, the female moves a foot or so upstream and digs another nest, throwing the gravel she digs up back over the first nest in order to protect it. This goes on until she has laid several thousand eggs (king salmon average 5600 eggs). The eggs are considered a delicacy by other fish. This is why they must be buried in deep holes. The Dolly Varden trout is especially fond of salmon eggs, and will follow the salmon long distances upstream in order to find their spawning grounds.

After the eggs are laid, Pacific salmon guard the nest for a short while, then relax and float downstream, tail first, dying within a few hours. Atlantic salmon do not die after spawning, but return to the ocean, as do steelhead trout.

Each of a salmon's scales has many rings on it that look like the rings in the trunk of a tree. The scales grow as the salmon grows. During summer, when food is abundant, the salmon grows swiftly, and so do the scales. Rings that are formed on the scales during summer are far apart. Rings that are formed in winter, when food is scarce, are close together, for during this time, the fish and its scales grow slowly. By counting the number of groups of close-together rings on a scale, one can tell through how many winters the fish has lived. One can tell something about how much food was available to the fish during each year of its life by noting during which years the scales grew most. It is also possible to tell just how old the fish was when it entered the ocean.

Types of Salmon

The Chinook or king salmon is the largest and finest type of salmon, sometimes weighing more than one hundred pounds, although this is rare today. Its flesh is deep red and very delicious. It lives from four to seven years before it spawns.

Silver or Coho salmon are not considered quite as desirable as kings, and are a bit cheaper in price. Silvers and kings are the types most often sold in fresh fish markets. The other species are usually canned.

More sockeye salmon are canned than any other type. They are especially plentiful in Alaskan waters. Sockeye are unusual in that their spawning grounds tend to be in streams which are connected to lakes. They seem to like the calm lake waters, whereas other species like the swift currents of streams. It seems that at times in the past, sockeyes have entered lakes, perhaps when streams were abnormally high, only to find themselves trapped when the waters receded. The sockeyes lived their lives in these lakes, and today, certain inland lakes with no access to the sea still breed these landlocked salmon. The main difference between them and their ocean brothers is that they are smaller, for salmon grow best in the ocean, where food is more abundant.

Dog salmon are also canned in great numbers. They received their name in the Arctic, where they used to be dried and fed to dog teams.

Humpbacked salmon are the smallest of the species, averaging about six pounds. Their flesh loses its firmness when canned, but they are canned nevertheless, for they are very abundant, and the better types are often scarce.

The steelhead trout is very closely related to the Pacific salmon, although it doesn't die after spawning. It is called a steelhead because of its very tough skull. It takes an extremely hard blow with a hammer or club to kill it.

Male salmon after spawning.

Sockeye salmon.

Humpback salmon.

Catching the Fish

The most efficient means of catching the salmon ever developed was the salmon trap. It was positioned across migration paths, and delivered fish of exceptionally high quality, for the salmon remained alive until they were removed from the trap. The initial cost of building a trap was high, but the price per pound was lower than any other method, for no fishing boat or other expensive gear was needed. It is believed that a relatively small number of traps placed in the right locations could capture most of the salmon that spawn in North American waters. It is for this reason that their use had to be strictly controlled.

The traps were large, and would hold scores, or hundreds, of fish at a time. They were made from walls of webbing. As the fish swam toward the trap, they first encountered a long wall of webbing that diverted them from their path through a series of V-shaped openings into the heart of the trap. The trap depended on the salmon's unwillingness to retrace its steps once it ran into an obstacle. The tendency of the salmon was to grope its way forward along the wall into the trap.

Because the traps were so effective, there were bitter disputes in the past over who had the right to put traps in the best locations. Trap sites were bought, sold and rented. Gradually, the rich and powerful canneries acquired most of the traps. This was terrible for the fishermen. The traps caught so many fish that the bargaining position of the independent fishermen became very weak, and the canneries were able to make them sell their fish at very low prices.

In 1935, traps were made illegal in Puget Sound. Alaska banned them in 1959, except for some Metlakatla Indians. California forbids salmon to be taken by any means except trolling.

In all areas of the West Coast except California, the purse seine and gillnet account for most of the catch. Most salmon taken in seines are canned, for they are often too beat up, and not cleaned and put on ice quickly enough, for the fresh or frozen fish markets. The advantage of seines is that they can take large numbers of fish in short amounts of time.

Some Puget Sound fishermen use the reef net, a type of gear not found elsewhere. It consists of a net suspended between two boats in a salmon migration path. When a sufficient number of salmon are taken, the fish are removed and the net is lifted, allowing salmon to pass underneath. Fish caught in these nets are of unusually high quality, since they are removed and put on ice soon after being caught, and since they sustain little damage from this type of net. Only a small percentage of the Puget Sound catch are landed in this way.

Trolling, employing four to eight lines with hooks attached, is used mostly for king and silver salmon. Trolling is the only widespread method of catching salmon where the fish are cleaned and washed immediately upon being

Alaska limit boat purse seining for salmon. *(Marco)*

caught, and iced soon after. Thus, fish caught by trollers command the highest prices. Most fresh and frozen salmon are caught by trollers.

Trolling has been called the "gentleman's business," not because it is easy (trolling crews work seventeen hours a day), but because it is less taxing on men and boats than some other types of fishing. It also requires more skill than perhaps any other type of fishing.

The Fine Art of Salmon Trolling

Salmon men have all kinds of methods of finding fish. One man described to me how he does it:

"You find what they're feeding on, and you'll find the salmon. Watch your sounder [a sonar depth recorder that gives a continuous graphical readout of depth]. It'll give a jump when you're over a school of herring. Herring is salmon food. Sink your lines smack into the middle of that school, and you'll get some salmon."

When your sounder jumps, how do you know it isn't just a big rock under the boat?

"If the jump is real steep, and maybe there's a few fringes on the side, then it's a school of fish. You need to know how to read your own sounder."

Do you use spoons and lures on your hooks, or do you use bait?

"Oh, I use hardware in the beginning of the season. Then I switch to bait. Herring or squid. You want to use something that will blend in pretty nicely with whatever it is the salmon are feeding on. If it's something dark, you might use squid. Other times, herring might be better. The color of the water gives you a clue what to use."

How can you tell when you get a bite?

"You keep an eye on the springs that hold the lines in place. When one stretches, you know something's bit. Many of us put bells on the springs, to make sure you notice when you got something. It's easy to miss if there's no bell, especially with small fish.

"You know, I can tell what's on the other end of the line before I pull it up. If it's a king salmon, he'll give a hard jerk, then wait a bit, then another jerk, and so on. Silvers will give rapid little jerks on the line. Lingcod will give a pull—not a jerk, but a pull—set there for a while, then give another pull.

"Some men haul their lines as soon as they get a bite. Others, like myself, let the fish "soak" for a spell. The fish's struggling attracts other fish. Also, it's best to let a big fish tire himself out before you try to haul him aboard.

"I'm mainly interested in salmon, but I'll keep lingcod, black cod and halibut if they happen to bite my hooks. A forty-pound lingcod brings around four dollars. There's not much of a market for hake or rockfish, so we throw them back.

"As soon as I get a salmon aboard, I kill it by clubbing it over the head. There's no point in letting it thrash about. It just rubs its scales off and bruises itself, which lowers its price."

(Rochelle Smith)

The Dory Fleets

Successful salmon boats are not always large, equipped with hydraulic gurdies and ten thousand dollars worth of electronics gear. Some are only eighteen or nineteen feet long and are powered by little outboard motors instead of three-hundred-horsepower diesels. In Oregon, especially, are found many of these dories. They have for years been trolling quite profitably for salmon. And no wonder, for they pay little or no dock fees and their initial cost and maintenance requirements are very, very low compared to the big boats. In areas where the salmon runs are close to a suitable launching spot, these craft are ideal. The most famous group of them is the Pacific City Dory Fleet in Oregon. They have been fishing for years with twenty-two foot plywood dories.

The dories are equipped with hand-turned instead of hydraulic gurdies to reel in the lines, and have centerboards instead of keels. They are trailered down to a beach near the fishing grounds and are launched right into the surf. The fishermen row past the breakers, then drop the centerboard and outboard motor in place.

Vanishing Salmon Runs

Once upon a time the Columbia River had the biggest salmon run in the world. That run is now destroyed, largely because of the numerous dams on the river. The dams prevent salmon from reaching their spawning grounds. When the salmon get stopped by a dam, they either don't spawn, or spawn in an unsuitable place and none of their eggs are able to hatch. There are many people fighting to prevent Canada's Fraser River runs from being destroyed like the Columbia's were.

The river started deteriorating in the 1930s with the building of the Bonneville Dam, 35 miles east of Portland.

(Washington Department of Fisheries)

The Grand Coulee Dam and the Chief Joseph Dam, built in north-central Washington, cut off two of the Columbia's finest runs, a run of giant June Hog salmon (so named because they begin their run in June, much earlier than most other salmon) and a run of sockeyes that was perhaps 1200 miles long, extending to the very headwaters of the river.

Newer dams employ fish ladders that are supposed to enable the salmon to climb over the dams and continue their journey. It has been found that only about half the salmon make it up each ladder. Thus, if there are ten dams the fish must cross, the salmon population is halved ten times, which means only about one in a thousand make it to their spawning grounds.

Dams kill salmon in other ways, too. When water spills over a dam, excessive amounts of nitrogen get into it. Salmon get nitrogen bubbles in their blood as they swim through this water. This is similar to the disease divers get called the "bends," and it kills many fish. Dams also slow down the flow of water. These turbid waters sit baking in the sun, and attain temperatures too high for the fish to survive.

There are many other ways salmon runs are destroyed. A century ago, gold hunters used a method called hydraulic mining, where tons of earth were washed away in order to expose veins of gold. This earth eventually found its way into streams, and the gravel beds of numerous spawning grounds were covered with mud. Salmon can't bury their eggs deep enough in mud to prevent them from being eaten by other fish. California's Sacramento River spawning grounds were ruined in this way.

Logging also decimates salmon populations. When slopes are clearcut, severe erosion takes place the following winter, and acres upon acres of silt are washed into the streams and onto spawning grounds.

Centuries ago, salmon used to be as numerous on the East Coast as on the West Coast. They were so numerous that early settlers used them for fertilizer. Damming, overfishing, logging and pollution wiped them out. In 1968 only about 500 salmon were taken from all the rivers in the state of Maine.

There are efforts to repair the ruined runs both on the East and West Coasts. A salmon hatchery built on the Elk River in Oregon significantly raised the salmon population on the river within three years after it had been built. Commercial fishermen were able to fish a full season, and caught quite a few thirty-pound three year olds that had been born in the hatchery. People feel that logging has been to blame for the destroyed spawning grounds on the Elk River. The hatcheries restore spawning grounds by removing the silt and mud with pumps, exposing the clean gravel the salmon need.

The History of Salmon Canning

Napoleon Bonaparte was partially responsible for the first canned food. He was having trouble feeding his armies, who were fighting on faraway battlefields. Hardtack, smoked fish and salted meat spoiled before it reached them. He offered a 1200-franc prize to the man who invented a better way of preserving food.

Nicolas Appert, a chef, pickler, preserver, winemaker and brewer, won Napoleon's prize in 1809 by developing a method of sealing food in jars and then boiling the jars. There were problems. Boiling water was not hot enough to kill some bacteria. Also, the glass jars often broke during shipment.

A few years later, Appert began using tin cannisters to seal his food in. He met with opposition, for many people felt that the tin would get into the food and poison people. His method caught on, though, and revolutionized the eating habits of the world. It was especially important to fishermen, for now their catch was able to be preserved indefinitely, and shipped anywhere in the world.

The first North American salmon cannery was built on the Bay of Fundy in eastern Canada in 1840. The first on the West Coast was built in 1864 by George Hume and Andrew Hapgood.

Hume arrived in California in 1852, during the Gold Rush. He came from Augusta, Maine, where his father and grandfather had both been fishermen. His ancestors in Scotland had fished salmon, from the Tweed and Tay Rivers. He caught silver salmon for a while on the Sacramento River and sold his catch to the miners and prospectors. In 1864 he brought Hapgood, his boyhood friend, out West and started a small cannery.

All of the cans were made by hand. The ends were cut out with a pair of shears, and stamped in a hand press. Sides were cut and then shaped on a reamer. Soldering irons, sitting in pots full of smoldering charcoal, were used to laboriously solder each can shut. About a hundred cans could be completed per day. After this first cannery became a success, thousands of men entered the canning business. Many were unsuccessful gold hunters.

Canning operations in Astoria, Oregon in the 1880s and 90s went something like this: A half-dozen boats were usually tied up at the cannery's dock on the Columbia River. The men in them, knee deep in salmon, swung them one by one onto the deck. Swarms of flies and hornets buzzed around, nibbling at the fish. It wasn't unusual for a boat to contain 200 fish.

The dock men slid the salmon down a chute from the dock into the fish house. The fish house was filled with Chinese workers, the flashing of long knives, and the smell of salt and cooking fish.

Two men caught the salmon at the bottom of the chute and flipped them over to the cutting table, where Chinese butchers, stripped to the waist, lopped off the heads, tails and fins, slit open the bellies and removed the entrails. After this Indian women washed them in large tanks.

Next, other Chinese men cut the fish into smaller pieces and stuffed them into cans. Fifteen thousand cases of Alaska Talls were stuffed and sealed each day by a good-sized cannery.

The cans were handmade, usually by Chinese during the off-season. After being soldered shut, the cans were dipped in brown lacquer to prevent rusting.

A Caucasian inspector tapped each can with a mallet to see if it was airtight. He could tell from the sound it made whether or not it was a leaker. About one in every dozen was a leaker. Leakers were thrown away to rot, for salmon was plentiful and cheap, and it wasn't worth spending time resoldering a can.

Many people disliked the Chinese, for they worked for lower wages than the whites would, and thus took jobs away from the whites. Whites tried various methods of getting rid of the Chinese. In 1904, E. A. Smith invented a machine that automatically cut open and cleaned salmon. In a few years, he had improved it to the point where it could handle eighty fish a minute, with less waste than when humans dressed the fish. In a few years, canneries from Alaska to California had replaced their human slitters with Smith's machine. They switched to Smith's machine both because of the higher profits that could be made using it, and because it gave them a way to get rid of many of their Chinese workers.

Lobster Fishing in New England

A Lobsterman's Day

The alarm goes off at 5:00 A.M., and I roll over and switch on the radio. The weather report says fog, but clearing by noon, so I decide to make a day of it. I wrench myself out of bed and pull on my clothes. My wife gets up and goes into the kitchen to make a quick breakfast. She works softly, so as not to wake the kids.

I'm out the door by 5:30. It's still pitch dark. I load my bait into the pickup and drive to the wharf. Damn, it's so foggy, I can't see to the end of the block.

My little skiff is waiting at the wharf. Even at this hour the wharf is bustling with men loading their boats and heading out for a day's work. I heave my bait into the skiff and row out to my boat, the *Lena,* tied to a buoy about fifty yards off the wharf. The *Lena* gently bobs up and down on the quiet waters of the bay like an oversized cork.

I tie the skiff to the buoy, hop on board the *Lena,* start her up and head out toward my strings of traps. Boats appear out of the fog and vanish into it like phantoms. I go slowly, because visibility is terrible. I'm not worried about finding my traps, though. I've been over every inch of this bay so many times, I could tell you within a gnat's ass where those pots sit.

The bay isn't too crowded with other boats as soon as I leave the harbor area, so I ease out the throttle a bit until I'm doing about six knots. I start up the bilge pumps, as the boat has taken a bit of water during the night, and fill up the live tank with seawater. Green's Island appears out of the fog dead ahead, silhouetted in the first light of the day. When I get to maybe a hundred feet off the shore, I turn the *Lena* and head her parallel to the shoreline, keeping a constant distance away and watching for my marker buoy. It bursts from the fog at about two o'clock off my bow, and I pull alongside, snag the buoy with a gaff, haul it aboard and slip the line that's attached to it around the pulley of the hydraulic pot hauler. The buoy is painted with the familiar green and yellow pattern that identifies it, and the trap that is attached to it is mine and only mine. Sometimes a guy'll "forget" what his buoys look like, and just happen to pull up someone else's. My friend, Al Pickering, caught a man doing this once. He came around a point and not thirty yards away some ragpicker up here on vacation was emptying one of Al's traps. Now Al's got one of the meanest tempers of any guy I know, and he's got the muscles to back it up, too. He boarded that ragpicker's boat so fast, the man didn't even have time to throw Al's buoy overboard. Al grabbed the guy with one hand and slammed him down on the deck, and waved his big hammy fist in front of the guy's nose. "If you ain't gone when I get back to the wharf," he told him, "I'm gonna pound you till there's so much blood, them sharks gonna jump right up on shore to get at you!" The man left our town that afternoon, and picked safer places to spend his vacation after that. And I don't blame Al a bit for the way he blew up. Lobsters are our way of life. If someone steals our catch, we go broke.

I fish singles—one trap attached to each buoy. My friend Ralph, in the *Sally and Jean,* he fishes doubles—two traps to a buoy. Some other men around here, they attach ten or so traps together and set them all at once. You can work more traps that way, but you can't cover as large an area. Which is best? Well, it all depends on conditions. If you've found some super rich grounds, ten traps might work fine. But on this bay, lobsters are gettin' pretty scarce, and the more ground you cover, the more lobsters you catch. That's why I wouldn't fish nothing but singles or doubles.

When the pot comes out of the water, I swing it onto the cap rail and open it. I take the legals out and dump them in the live tank. Shorts have to be thrown back, 'cause there's a heavy fine if I'm caught with one. Sometimes rock crabs and starfish come up in the traps. I return them to the sea, because they're not worth anything. I plug the claws of the legal-sized lobsters with wood plugs, to keep them from tearing up the others in the tank. Lobsters might not look it, but they're savage creatures. They'll fight to the death, and then eat their opponent after they've killed him.

I replace the used bait in the trap with fresh stuff, repair any minor damage, like a broken slat or two, throw the trap and buoy back in the water, and go on to the next buoy.

The first trap I just finished with had half a dozen legals in it—not bad. Three of the legals are selects—pound and a halfers that bring the highest prices. Lobsters of any other size aren't worth as much. Near as I can make out, it's because they don't fit nicely on the dinner plates of them fancy Manhattan restaurants.

About eleven o'clock I decide it's time for a break. The fog has lifted, and I see my friend Henry in the *Sea Wolf* not half a mile away. I chug on over, and we drink a couple of beers and eat our lunches. It's turned into a calm, sunny day, and we're both enjoying it. Yesterday, the wind was whistling like a banshee and whipping up huge waves, and I was forced back to port early. Storms can start in a matter of minutes, and small boats like mine get swamped at that time. Each winter, one or two guys I know or have heard of get lost while at sea.

The afternoon progresses like the morning—pull alongside a buoy, gaff it and haul the trap, empty and repair the trap, throw it overboard and go on to the next one. I work 450 pots myself, which keeps me busy seven days a week and eight months of the year, weather permitting. During the winter months, I crew on board a longliner, out for bottomfish. Lena, my wife, wishes I'd stay home more.

I finish with the last of the day's traps and head the *Lena* home. I stop first at the buyer's. Just my luck! Charlie, the owner, is working today, and he is as tight-assed with his money as they come, always looking for some excuse to knock down his price. One broken claw, and he'll cut the price on your whole day's catch, and no one will make out, no one but Charlie, that is. After he helps me unload, I refuel my boat, and the cost of the fuel is subtracted from what Charlie owes me for my day's catch. Then I wash down my boat with seawater, repair a few of the badly damaged traps I didn't have time for earlier, and go over to Henry's boat again and drink a few beers before returning home. We gab about the day's catch, complain about the Japs and Russians taking all our fish, and talk about how every year, fewer lobsters are caught in our bay. Not many men get rich lobstering, but I love it. The wind, the sea, they're part of me. As long as I'm still able to walk, you'll find me out on this bay, hauling pots.

History

Americans have caught and eaten lobster since the time of the first New England settlements. Accounts as early as 1605 refer to the great abundance of lobsters in New England waters, and speculate on the large profits that could be made from them. Captain John Smith wrote in 1616, in his *Description of New England,* that there was hardly a bay, shallow shore or cove where as many clams and lobsters as one wanted could not be taken. Storms often washed large piles ashore, making it possible to gather them without even getting one's feet wet. There are accounts of lobsters six feet in length being found!

During this period of early colonization, most families lived close to the ocean, and were able to gather all the lobsters they needed. Thus, there wasn't any point in trying to sell them. It wasn't until people settled in the inland areas that the market for lobsters arose. By 1740 lobster peddlers were a common sight in Boston. Prices were probably quite low, due to the great abundance. Old men and young boys most likely did the lobstering, for fishermen in their prime were busy working the more profitable cod fisheries. In those days, lobsters were caught with gaffs and spears, or by simply wading into the water and picking them off the bottom with one's hands.

By 1790 lobstering had grown to the point where there were five vessels regularly carrying the crustaceans from the tip of Cape Cod to New York. Commercial lobstering began in Connecticut in about 1810.

For two centuries, men took as many lobsters as they pleased, for the stocks seemed limitless. Finally there were signs of depletion, and in 1812, Provincetown put into law the first restrictions on the fishery. The law forbade nonresidents of Massachusetts from catching lobsters off of Provincetown. Lobster stocks continued to be depleted, though, and in the 1830s Boston sent a fleet of well smacks (large transport ships for lobsters and other sea life) to Maine in search of virgin grounds.

The hoop net, or plumper, came into use in the 1840s and 50s. It was made of an iron or wooden hoop, across which netting was strung. The hoops were set on the sea bottom with bait and haul lines attached. A lobster would walk into the hoop and nibble at the bait. The idea was to haul the net up to the surface before it finished eating and walked out. To accomplish this, plumpers had to be hauled and checked every quarter of an hour or so. The lobsters were kept in "live cars," just as they are today. Live cars are boxes constructed of planks and placed underwater. The planks are far enough apart to allow seawater to circulate and keep the lobsters alive. At the end of the day the live cars are towed to market or emptied into a well smack.

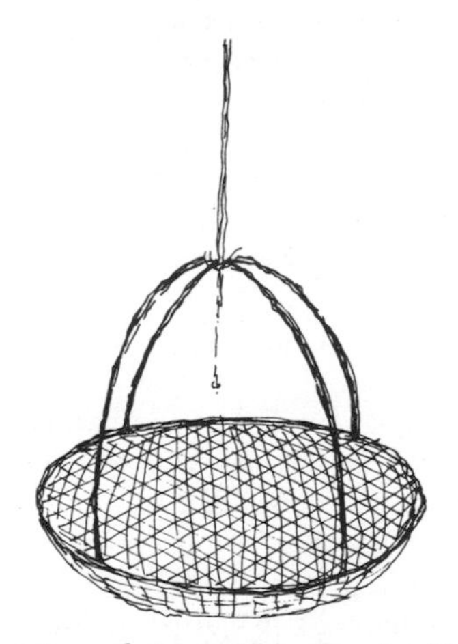

Hoop net, or plumper. *(John Douglas Moran)*

Wire-mesh lobster trap.

Lobster canning began in Maine in the 1840s. By the 1880s, twenty-three canneries were operating. Not one of them, though, was in Massachusetts. The demand for fresh lobsters in Massachusetts was so great that there were rarely any left over to be canned.

Traps came into common use about a century ago. Lobstermen set them over large areas, and it was soon found that rowboats were not speedy enough to tend all of them. Sailing vessels began replacing the rowboats. The most popular vessel was a double-ended dory fitted with one or two masts and a centerboard. It required a two-man crew—one man to handle the boat, and one to haul pots.

By the turn of the century, lobsters were being shipped all over the country in refrigerated railroad cars. Seawater was dumped on them every few hours to keep them alive. About half of them would survive a coast-to-coast trip.

The last half-century has seen several technolgical changes in lobstering that have eliminated much of the manual labor. Between the world wars, inexpensive engines were installed in the sailing craft. As the Second World War approached, boats specially designed for lobstering began to appear. Hydraulic and mechanical pot haulers were introduced around this time and lobstermen were saved the backbreaking work of hauling pots by hand.

The motorized vessels enabled fishermen to greatly increase the number of traps they could work. This quickly depleted lobster stocks,which forced fishermen to set even more traps. There are many who feel that this overfishing will soon ruin the inshore lobster grounds. Nearly ten times as many lobstermen fish Massachusetts waters as did in 1830, and yet the total catch has only doubled since that year.

The Lobsters

The world-famous Maine lobster is actually *Homarus americanus,* the American lobster, and is found from Labrador down to North Carolina. It is found in its greatest abundance off Maine and Nova Scotia, in the Gulf of St. Lawrence, and on Georges Bank, off Massachusetts.

The lobster is a cold-blooded sea creature of the class Crustacea, meaning "crusty," or having a shell. Shrimp, crabs, crayfish and barnacles are also members of this class.

A lobster's head and thorax are contained in its main part, called its body. Its abdomen is located in its six-jointed tail. A lobster has very poor eyesight, and is probably able to distinguish little more than light from dark. Its eyes are mounted on swiveling stalks. Touch, taste and smell are its best-developed senses, and are incorporated into a system of tiny hairs distributed over all of its body. Lobsters also have two pairs of antennae or "feeders," to help them find their way. Lobsters do not see a lobster trap, but either bump into it or are drawn to it by the smell of the bait. They have no ears, but can probably detect vibrations in the water through their sense of touch. Loud noises seem to scare them away, for lobster catches are poor in areas where there has just been a fireworks display.

Lobsters seek out rocky, kelp covered bottoms that offer lots of hiding places. Since they are cold-blooded, they are greatly influenced by water temperature. It affects their rate of growth, their metabolism, and the time of year that they molt (shed their shells). They can live in waters as cold as 29 degrees Fahrenheit, or as warm as 90 degrees, but a quick *change* of 25 to 30 degrees will kill them. Many boats fishing in deep waters run into this problem, for the temperature difference between the bottom, where the lobsters are caught, and the surface, where the boats are, can be as much as 50 degrees.

Warm waters make lobsters active and hungry, thus increasing the chances that they will enter a trap to feed on the bait. Lobsters like water with a salt content of 3 to 3½ percent. Fewer lobsters are caught near river mouths, sewage outlets or storm drains, where fresh waters mingle with the seawater and dilute it to below its optimum level of salinity.

Lobstermen have many opinions on which is the best bait for use in the traps, although they do agree that fish are better than crustaceans or mollusks. Redfish, flounder and mackerel are often used. In their natural habitat, lobsters scavenge the seafloor for parts of freshly killed fish. When they don't find any, they kill and eat clams, mussels and other lobsters. They will even eat a freshly killed seagull, but they will not touch putrid, decaying flesh. When a lobster has an excess amount of food, he often buries it like a dog with a bone, and chases other lobsters away from the spot.

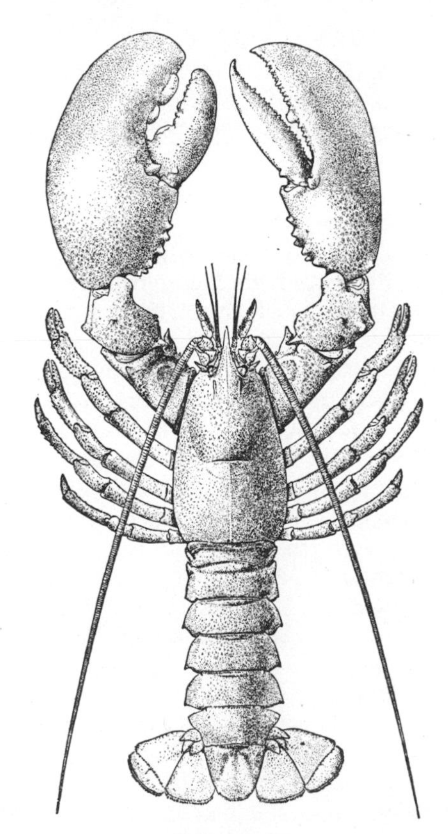

Male lobster.

(Maine Department of Marine Resources)

A lobster is only a bit heavier than an equal volume of seawater. (This is called "slight negative buoyancy.") Since it is almost but not quite buoyant, it is able to walk around on the ocean bottom effortlessly, on the tips of its thin legs. Lobsters forage at night mostly, and rest during the day, which is why traps left out overnight catch more than those left only for the day. The lobsters seem to prefer dark areas over light, probably because they feel safer. In captivity, they seek out and huddle in the darkest areas of the enclosure. In shallow ocean waters, where sunlight is able to filter down to the bottom, they are most often found under rocks or piles of kelp, or in other areas shielded from the light. It has been found that if old automobile tires are placed on the bottom, lobsters will enter and hide in their dark recesses.

A lobster must shed its shell from time to time, for its shell is rigid and is not able to grow with the rest of its body. In order to crawl out of its old shell, it must compress its claws to a quarter of their normal size. This is why the claws of a lobster that has just molted are wrinkled and distorted. The lobster absorbs considerable amounts of water soon after molting, and bloats its claws back to their original size. Some of the water is absorbed by the rest of the body, making its meat watery and not as tasty as usual. These watery-fleshed lobsters are called "jellyrolls," and don't bring as high prices as those with firm, hard shells, for hard shells are a sign that the lobster has not molted recently.

Inshore Lobstering

Inshore lobstering (lobstering close to shore) is done using baited traps, or "pots." They are made by nailing spruce slats over oak frames, and are little changed from pots used a century ago. The wood is untreated, for lobsters can smell if preservatives have been painted on the wood, and tend to avoid such traps.

Bricks or pieces of cement are put in the pots to keep them on the bottom. Each pot is connected with a nylon line to a plastic buoy on the surface. Sometimes empty plastic bleach bottles are used for buoys.

New England lobstering is done mostly in summer, for lobsters are scarce in winter, and the weather is fierce and dangerous. Some say that lobsters move offshore in the winter, when the shallow inshore waters become too cold and turbulent for them. Others say they go into a state of dormancy, like blue crabs. The lobstermen spend their winters repairing gear, taking vacations, or setlining for cod and haddock.

Lobstering is a hard way of life, with long hours of work under fairly dangerous conditions. Few young men are entering the fishery these days. A study of the Massachusetts area revealed that less than 20 percent of the full-time lobstermen were in their twenties. The average age was over forty. The study also revealed that the average educational level of a lobsterman is higher than in other fisheries. Most have high school diplomas, and some hold college degrees. Virtually all of the lobstermen are Caucasian, married and have children. Many come from old New England families, and most have spent their lives in the fishery, beginning as kids working on the family boat, or as "coolies" (hired help) working on someone else's boat. The vast majority interviewed in the study wanted no other job, and were happy with their independent way of life.

Lobstering Territories

Summer vacationers in Maine often try their hand at lobstering, hoping to pick up a few extra dollars. They'll buy a whole set of equipment—traps, lines and buoys—and set it in a likely looking spot near shore. They might fish for a week or two, take a few lobsters, but one day they'll go to check their traps and find that someone has cut their lines, and all the fancy gear they just bought is lost. They might complain to the sheriff or ask around town, but the culprit probably won't be caught, for what he has done has the tacit approval of the community.

Maine lobster territories are the most jealously guarded fisheries in the United States. Each harbor "owns" a certain area around it, and even men from neighboring harbors will have their gear destroyed if they fish there. The right to fish in a harbor's territory is a privilege that is handed down from father to son.

Harbor "gangs" are called by the names of their harbors: for instance, the "Monhegan boys," the "men from Boothbay," or the "Friendship gang." Harbor gangs are tightly knit fraternities of men, most of whom have grown up together, and it is very hard to be admitted to one of these gangs. It helps to have family members who are long-term residents of the harbor town you want to fish out of. It helps to start young, too, like when you're fourteen, working a few traps out of a rowboat after school. If a boy starts very young, he has plenty of time to learn the unwritten rules and customs of the gang. But if he is just a summer resident of the area, and comes from an out-of-state, nonfishing family, his chances of joining a gang are slim indeed.

Territories overlap, especially far from shore, where boundaries are rather unclear. It is easy for a lobsterman to stray into another territory. Sometimes this sets up a whole series of line cuttings and retaliations, and puts several men out of business. There are cases, although not common, of two gangs completely destroying each other's gear and then burning each other's docks.

Lobstermen don't like to destroy another man's gear. They do it secretly, and won't talk about it afterwards. But they feel they have to protect their own livelihoods. "Those harbors that allow no outsiders are the ones with the best catches, year after year," says one old-timer. There aren't all that many lobsters in New England coastal waters anymore, after 350 years of heavy fishing. New England lobstermen need to protect this vanishing resource, or they'll be out of jobs.

Waters around some of the small, privately owned islands that dot Maine's coastline are considered exclusive territories of the families that have owned those islands for generations, even though by law the waters are in the public domain. If the families don't have enough relatives to work the grounds, they will rent out the right to lobster there. Certain families have owned the rental rights to some of these grounds for generations. These private islands aren't fished as heavily as mainland waters, and average higher catches.

Ragpickers

There is an especially strong prejudice against "ragpickers"—men who lobster part time, and also work at another job. As one lobsterman says, "They got a regular paycheck coming in each month, but they still want to take the bread and butter off our tables."

In Massachusetts, ragpickers are not dealt with as severely as in Maine and as a result, the coastal fisheries are in bad shape. In Marblehead, Massachusetts, there are only 18 full-time lobstermen, but there are 120 ragpickers. Lobster catches are steadily decreasing in Marblehead, for there just aren't enough to go around. Some ragpickers are teenagers after a little extra money, while others are school teachers who lobster during their vacations. Some view it merely as a hobby, while others need it to augment their income. Until 1971, Massachusetts ragpickers fished anywhere from 5 to 150 traps. Lobstermen were afraid the grounds would soon be exhausted, and got a law passed limiting the ragpickers to 25 traps and a four-month season. Full-time lobstermen claim that ragpickers are only after the quick buck, and don't give a hoot if the fishery is ruined. For instance, lobstermen avoid setting pots in harbor regions, in order to give the lobsters a needed refuge during breeding and molting seasons. Ragpickers ignore this custom, and set their traps all over the harbor.

The most hated of all ragpickers are the scuba divers. Lobstermen believe that the divers get their lobsters by stealing from traps.

There is talk of raising the price of a commercial license to $500. This high fee would make it unprofitable to lobster part time. The full timers would catch the lobsters the ragpickers are now getting, and would thus come out ahead, in spite of the stiff licensing fee. There is also talk of doing as the fishermen in Maine do, and destroying ragpickers' gear.

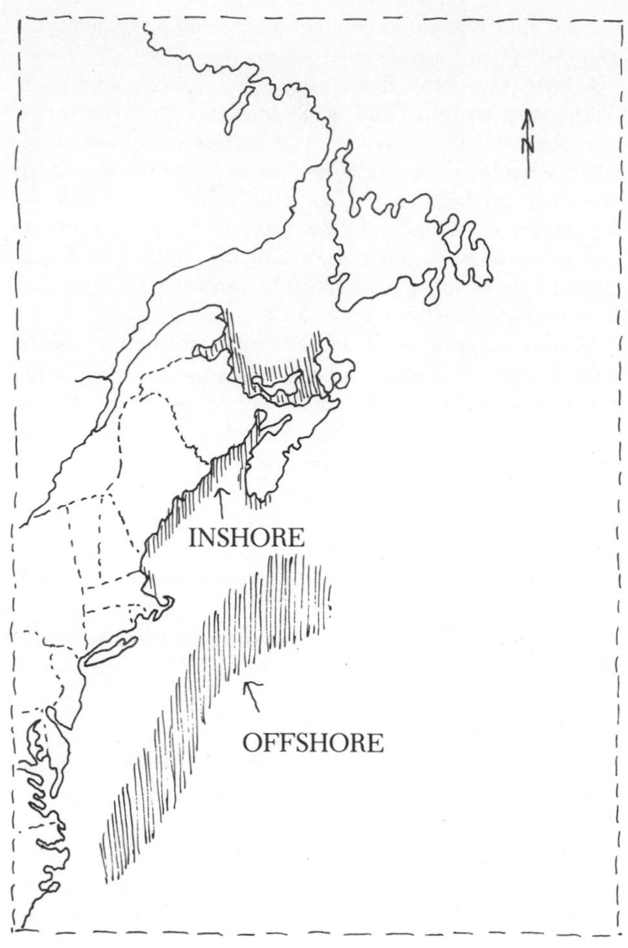

Lobster fisheries. *(John Douglas Moran)*

Doing It the Old Way—Oar Power

Henry is not your average commercial lobsterman. He does not own a $15,000, high-powered Jonesporter, but instead fishes out of a small double-ender that he rows by hand. He hauls his traps by hand, too, instead of using a hydraulic pot hauler like the power boats do. And he has been fishing this way for sixty-three of his eighty-one years. His doctors tell him rowing is good for him, and to keep it up. He can't think of any reason why he wouldn't want to, as long as his body holds up. He doesn't catch as many lobsters as the power boats do, but he sure has them beat when it comes to operating expenses. He doesn't have to worry about the rising cost of gasoline and diesel oil, or about a broken pot hauler or bent propeller. He gets up before dawn every day, goes down to his boat, harbored at Vinalhaven, Maine, and is out on the water in time to see the sunrise. He loves his work, loves the early morning, and loves observing all of nature's wonders. He believes that people who rise early are more optimistic about life.

Henry tells many stories about unusual things that happened to him while lobstering. One is about a seagull who stole a lobster from him. He was rowing his double-ender one August morning by Green's Island shore. His lobsters were sitting behind him in a tub. One wise seagull realized that Henry couldn't see him as long as he stayed behind Henry's back. The gull zipped down and grabbed a two-pound lobster from the tub before Henry could turn and swat him. He gripped the lobster in just the right place so that it couldn't grab a hold of him and pinch him. He landed on Green's Island, laid the lobster down, reared back and drove his beak right through its shell, after which he ate it at his leisure.

On another occasion, Henry pulled up a lobster pot, and inside was a 2½-foot catfish, ugly as all get out and with teeth like a bulldog. Henry was in a quandary, for he didn't know how to take it out without getting his hand bitten off. Finally he took his gaff, stuck it into the fish's mouth, and pulled him out of there. He couldn't figure out how to get it off the gaff, so he just threw the whole mess overboard, gaff and all. The last he saw of the fish, it was swimming angrily away, the gaff trailing behind it.

If a man enjoys his work as Henry does, it doesn't matter that he spends a few more hours at it each day than do other people.

Trap Day at Monhegan Island

Monhegan Island is a mile-and-a-half-long piece of granite off the coast of Maine. It is only during the summer that the crowds of vacationers come to the island. After Labor Day, the small village of permanent residents begins preparing for lobster season, still four months away. A man loses as much as 100 traps during the season, and each one has to be replaced. The wooden traps are made by hand by the fisherman or his family. It takes about three hours and ten dollars worth of material to build a trap.

The traps are all completed by the middle of December, and are hauled to the docks. Lots are drawn to see which men get the choicest mooring spots. The best spots are on the edge of the dock, where pots can be quickly loaded onto a boat. A spot at the center of the dock is much less desirable, for every minute saved on the first day of the season is valuable. The lobsters have been left alone for six months, and are more numerous than they will be later in the season. A day of trapping in January is often worth a week in May.

The season begins sometime in January, when it looks like there will be several days of good weather. If a storm occurs on the day after trap day, disastrous losses in gear could occur. If it looks like a streak of decent weather is approaching, a meeting is called, usually by the fishermen most anxious to get things going. The word is passed from shop to shop and from house to house, announcing the meeting. After the fishermen gather, everyone expresses how he feels about starting the season the next day and a sort of vote is taken.

School is cancelled on the day, and the whole family pitches in and helps load its boats with traps. Hired help from the mainland is also used. The loaded boats race out of the harbor, set their traps, and return for another load. As many as fifty traps are loaded onto a boat at one time. The men try to set as many pots as possible before sundown. Tens of thousands of the traps are set on trap day.

Trap day at Monhegan Island. *(Maine Department of Marine Resources)*

The Lobster Buyer

Why will a lobsterman sell to the same dealer over and over again? Jasper, a small Maine lobster buyer, gives three reasons why his customers keep returning to him. The first is that he runs a friendly, informal business. The men come to hang out and shoot the breeze, as well as sell lobsters. Secondly, he offers them a workshop and an area where they can salt their bait. Also, he offers them bait at cost if they help him unload the bait truck. And thirdly, he owns some of the men's boats, and they'd darn well better sell to him, or they'd find themselves swimming out to their traps.

His dealer's shed is a small house with old notices, advertisements, government regulations, nautical maps and marine supplies slapped on the wall and flung about the floor. An old fish scale hangs from the ceiling. There's a stove and a few chairs strewn around, a dog sleeping by the stove and a few old lobstermen occupying the chairs. Jasper claims that the men feel comfortable with the informality of his place.

When a boat finishes for the day and pulls alongside Jasper's dock, Jasper and the fisherman unload the barrels of lobsters and wheel them over to his scale. He puts all the barrels on the scale, if they'll fit, and writes down their weight. Then he dumps the lobsters into a bin, takes

Hauling a lobster trap. *(Maine Department of Marine Resources)*

the empty barrels, weighs them and subtracts this weight from the first figure. He multiplies the weight by cost per pound of the lobsters, and then subtracts the cost of the gasoline that the boat needed. Then he gets out his money pouch and pays the man on the spot. Jasper feels that maybe he should only pay the men once a week instead of every day, for he has this notion that if you give a man a whole bunch of money at one time, he's just liable to pay you a bit of what he owes you.

The Offshore Fishery

In the last two decades, rich offshore grounds have been developed, and the lifestyles of many lobstermen have been drastically changed. They can no longer come home to their families each night, for the grounds are too far away. Some of the younger men have sold their boats, given up their independent way of life, and now work as crewmen on 100-foot lobster "tankers" that make two-week trips to the grounds. Most of the older lobstermen, though, value being captains of their own boats more than they do the high pay they could receive on the tankers.

The great abundance of lobsters in the offshore grounds is reminiscent of stories of the way things were on the New England coast 200 years ago, when lobsters were so plentiful, they were used as fertilizer. The question is, how long will it be before the offshore grounds are as depleted as the inshore ones?

Cod, Halibut and Other Groundfish

The Biology of the Cod

The cod is a member of a family called the Gadidae that includes the haddock, pollock and hake. Cod are caught as far south as North Carolina, but are found in their greatest numbers off New England and eastern Canada, in such places as Newfoundland's Grand Banks, Massachusetts's Georges Bank, or the Nova Scotian shelf. A small percentage of the North American catch comes from Pacific waters, although many of the species marketed as cod on the West Coast, like rock cod or black cod, are not cod at all, but belong to other families of groundfish. The term groundfish refers to those species of fish that spend their lives on or near the ocean floor. Adult cod live near the sea floor in water as deep as 1500 feet and as shallow as 18 feet. The larger varieties of cod weigh twenty to thirty-five pounds. The smaller weigh around twelve pounds. The largest known cod—the Patriarch Cod—was caught in 1895 off of Massachusetts and weighed 211½ pounds. It was over six feet long and possibly fifty years old.

Cod spawn in wintertime, in cold water at temperatures as low as 32 degrees Fahrenheit. The newly spawned eggs expand and rise to the surface, where they hatch. The cod larvae, a fraction of an inch in diameter, feed on the larvae of barnacles, shrimp, lobsters and crabs. If every egg spawned by every female codfish in one season were to hatch, and if each of the larvae were to grow to maturity, the oceans of the world would be filled solid with codfish. A large female spawns as many as nine million eggs at a time. Only one in a million, on the average, ever reaches maturity.

The fry sink to the bottom when they are two months old and spend the rest of their lives there. In deep water, they sink slowly, taking about two weeks to reach bottom, for at each new level they must acclimatize themselves to the pressure. Many fry die at these middle depths due to the lack of food. When they finally do reach the bottom, they must learn to prey upon quite different sea creatures than they ate on the surface. Their diet will consist largely of herring, menhaden and shellfish. One of their favorite foods is squid.

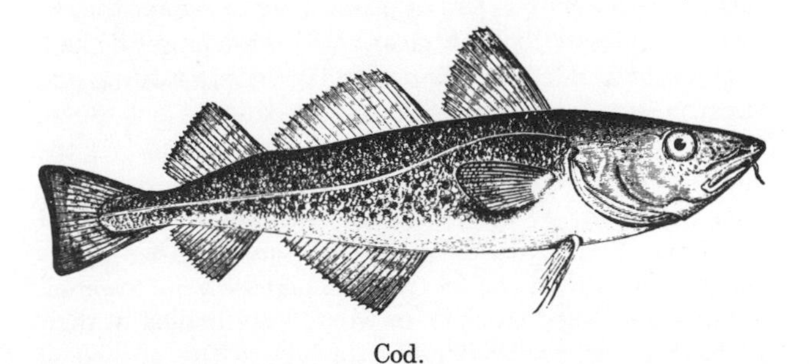

Cod.

Fishermen are very much aware of the cod's fondness for squid. In the old days, fishermen would spend considerable time before a season jiggin', or handlining, for squid, to be used as bait once the season opened. Arthur Scammel's poem, "The Squid Jiggin' Ground," describes a Newfoundland harbor scene sometime around the turn of the century:

> This is the place where the fishermen gather,
> In oilskins and boots and Cape Anns battened down.
> All sizes of figgers with squid lines and jiggers
> They congregate here on the Squid Jiggin' Ground.
> Some are workin' the jiggers while others are yarning
> There's some standing up and there's more lying down,
> While all kinds of fun, jokes and tricks are begun
> As they wait for the squid on the Squid Jiggin' Ground.

Further verses talk about some of the more colorful fishermen like Uncle Bob Hawkins who wears "six pairs of stockings" or old Uncle Billie, who gets so into his work that his "whiskers are splattered wi' spots of the squid juice that's flying around." The poem ends by concluding that squid jigging is hard work, man's work, and definitely not for greenhorns.

Scientists used to use the cod to learn about other species of fish, particularly mollusks. Before scientists developed the dredges that could sample mollusk populations on the ocean floor, they used to study them by examining cod stomachs for remains of the mollusks they had eaten. By recording the depth and location at which the cod were caught, and by assuming that the cod ate the mollusks shortly before they were caught, scientists could map out areas and depths that mollusks frequented. The first known specimen of the giant deep-sea scallop was discovered in 1845 by examining the stomach of a cod caught off of Maine. People soon realized that the scallop was quite delicious when fried, and in a few years, a sizable scallop fishery sprang up off of New England and

Baiting the hooks.

Setting the lines.

the mid-Atlantic states. Today, the waters off of New Bedford, Massachusetts are still abundant producers of scallops.

The Golden Triangle

Codfish have fed mankind for thousands of years. Ancient peoples used to preserve the fish by stretching them on short sticks and drying them in the sun, and it was from these short sticks that cod got its name. The stick was called a *cad*, or *gad*, an ancient Sanskrit word meaning rod. The Romans called the fish *gadus*. Several English words are derived from the same root: goad and, perhaps, cat (gad)-o'-nine-tails.

Cod have been fished commercially on this continent for more years than any other sea creature. Fishing for cod began soon after the first colonies were established. The Massachusetts Bay Colony was the first to market the fish on a large scale, exporting 300,000 dried salt cod to Europe in 1691, its first year of existence. Religious freedom is often cited as the main reason early settlers came to America. Another reason, less idealistic but certainly very practical, was the large profit one could make in the New World's abundant fisheries. Stories of the unbounded numbers of cod and other fish lured many settlers to America.

As the slave trade between the New World and Africa grew, a trade route developed called the Golden Triangle. A ship sailing this route left Boston, Marblehead or Salem loaded with salt cod caught in the New England area. The ship sailed to Spain or Portugal and sold its cargo. Then it swung down to West Africa and bought slaves with the money from the cod. It left Africa and sailed the Middle Passage to the West Indies where it unloaded the slaves and bought sugar and molasses for the New England rum distilleries. Finally it returned to New England, sold the sugar and molasses, filled its hold with salt cod, and began the cycle again. Handsome profits were made on each leg of the journey. Nothing was wasted, not even the partially spoiled or oversalted cod that was unfit for the Spanish and Portuguese markets. It was used to keep the slaves alive during the long journey across the Atlantic.

This method of earning a living sounds like one fit only for pirates or scoundrels, and yet many of today's upper-class New England families can trace their fortunes back to the salt cod, rum and slave trade of the last century.

Dorymen

Skippers of the 1800s had very few instruments to help them find their way. They would sail from port at a certain compass heading and would try to estimate, from the winds and from the time spent sailing, when they reached a known cod fishery, such as Georges Bank or Le Have Bank. When the captain felt they were getting close, he would stuff a glob of lard or tallow into the indented bottom of his sounding lead and heave it over the side. When the lead hit the sea floor, a bit of it would stick to the lard. The captain hauled up the lead and told by what was embedded in the lard what type of bottom he was sailing over. If the bottom was muddy, he'd sail on, for that meant there were few fish or poor-quality ones like the red hake. A fine sand bottom meant that there might be haddock around. But haddock don't take well to salting, and thus these too were bypassed by the old-time fishermen, for salting was their only means of preservation. If the sounding lead indicated they were over a coarse sand bottom, preferably with bits of shell in it, the captain would order handlines lowered, baited with squid and herring. If they came up with cod on them, the dories were lowered.

Dorymen fished alone or in pairs. They hauled in their 3000-foot setlines by hand. Attached to the myriad of

Taking in the fish.

Cleaning the fish.

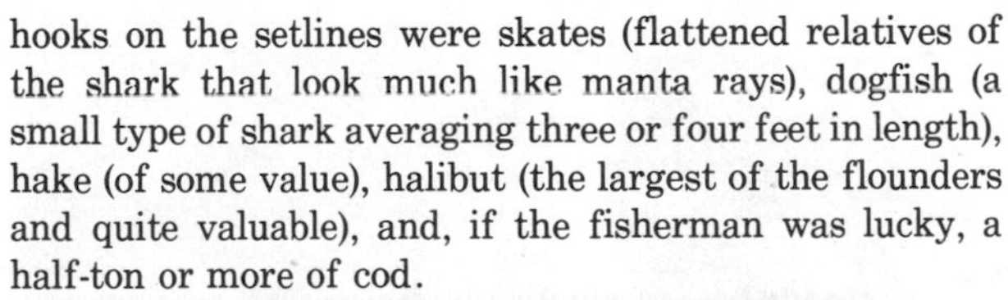

hooks on the setlines were skates (flattened relatives of the shark that look much like manta rays), dogfish (a small type of shark averaging three or four feet in length), hake (of some value), halibut (the largest of the flounders and quite valuable), and, if the fisherman was lucky, a half-ton or more of cod.

The old dorymen used to land their fish without ever touching them with their hands. As they pulled them alongside the dory, they slit their throats to bleed them. After this, they snagged them with a gaff, extricated the hook from the fish's mouth by skillfully maneuvering the fish with the gaff, and flipped the fish aboard. When the dory returned to the mother ship the fish were tossed aboard using a long-handled pitchfork. At this time, if the weather was good, dressing and salting began. These operations were also done by dorymen, and this was the first time they actually laid hands on the fish.

The first man in the dressing crew lifted each fish by sticking one finger in its eye and another in its gill. He slit the belly open, removed the liver and tossed it in a bucket (cod livers were used for medicine). Another man beheaded the fish and scooped its guts out. A third man tore out its backbone and threw the gutted fish into a tub, where it was scrubbed clean of all blood and entrails. After this, it was tossed to the salting crew waiting in the hold. They rubbed coarse salt into the wet fish and stacked them as if they were pieces of cordwood. Salting took a certain amount of skill, for if too much salt was added the fish became tough as shoe leather, and if too little was rubbed in, the fish rotted.

Albert Jensen, in his book *The Cod,* graphically records the sounds of the different dressing operations. There was a tearing sound as the fish was slit open, a "carraach" as the backbone was ripped out, and various plops as the viscera and head were thrown on the deck. There was the "clik-nik" of knives, and the raspy sound of coarse salt being rubbed into flesh—a sound that resembled the whirring of a grindstone.

Unpacking the fish.

A boat headed for home after it had "wet all its salt." First, though, it had to visit all the nearby fishing schooners and take letters from the men for their wives, for some of the other boats would not see port again for several weeks.

Hih! Yih! Yoho! Send your letters raound!
All our salt is wetted, and the anchor's off the graound!
Bend, oh, bend your mains'l, we're back to Yankeeland—
With fifteen hunder' quintal,
An' fifteen hunder' quintal,
'Twix old Queereau an' Grand.

Dorymen's lives were hard, monotonous and sometimes quite dangerous. One of the worst things that could happen to them was to get caught in the "gray terror"—the summer fog. The Grand Banks of Newfoundland are especially known for this hazard. It would come up and engulf a dory swiftly and unexpectedly. The mother ship and other dories would vanish, and the doryman would start rowing toward where he thought the ship was. Now and then he would blow a note on his conch shell and if he was lucky, the schooner or another dory would hear him and answer. He could sometimes use these sounds to guide him home. There are many, many tales of dorymen not making it home. Sometimes their dories would be found days later, empty. Sometimes no trace of dory or fisherman would ever be found.

In 1933 the four-masted dory schooner *Sophie Christenson,* on a five-month trip to Alaska's Bering Sea, landed more cod than perhaps any vessel has ever landed on a single trip. It took 453,356 fish, which added up to about 700 tons. One doryman in the crew caught 25,487 fish, also probably a record. Another doryman caught 1051 fish in a single day. On the best day of the trip, the ship landed 16,851 fish. The ship suffered only one casualty during the trip. Dory number 13 got lost in a gale and was found several days later—empty.

Modern Methods of Cod Fishing

Dory fishing is no longer conducted by the American or Canadian fleets, although the Portuguese Grand Banks fleet still engages in it. In New England waters, thirty-five-foot power boats have replaced the dories. On the West Coast, diesel-powered halibut schooners and Pacific combination boats do the setlining.

In 1878 fishermen tried the Norwegian method of gillnetting for cod. It was particularly useful when the cod were spawning, for at that time they don't feed and thus have no interest in baited hooks. The gillnets were strung across spawning paths and the fish swam into them, entangling themselves. Today, gillnetting is rarely used to take cod.

The beam trawl net, a forerunner of the modern otter trawl, was introduced into North American fisheries in 1891. Beam trawling is an old European method of fishing that was used as early as 1376 on the Thames River in England. It is a fairly efficient method in areas where there are dense concentrations of fish. In other areas, though, the small net mouth severely cuts down on the number of fish caught. The otter trawl has a much larger mouth and is considerably easier and safer to use. It has almost replaced the beam trawl in North America.

In 1905 ships began using ice instead of salt to preserve cod. Soon after, it was discovered that hake and haddock kept well in ice, and their value skyrocketed. Before icing, they were next to worthless, for they cannot be preserved well by salting.

The public preferred the taste of fresh, iced fish far better than that of salted and dried cod, and icing soon replaced salting. In a few years, haddock, a close relative of the codfish, equaled and then surpassed cod in popularity. They were easier to catch, too, for the smooth sandy bottoms that were their habitat did not rip up the otter trawls as badly as did the rough bottoms the cod frequented.

Today's method of dressing the catch is a little different than it was during the days of dory fishing, in that after the fish are washed they are packed between layers of flaked ice instead of salt. Many of the more modern vessels have quick freezers and refrigerated holds aboard. A few, like the factory trawler *Seafreeze Atlantic*, are equipped with automatic gutting, washing and filleting machines.

Other Groundfish

Groundfish (fish that spend most of their time on or near the ocean's bottom) can be divided into three subgroups according to their shapes and habits: roundfish, flatfish and rockfish. Roundfish get their name because cross-sections of their bodies are fairly round, at least compared to those of the flatfish. Examples of roundfish are cod, sablefish (black cod), lingcod, haddock, hake and pollock.

Flatfish have flat bodies with both eyes on the uppermost side of the body. Examples are sole and flounder.

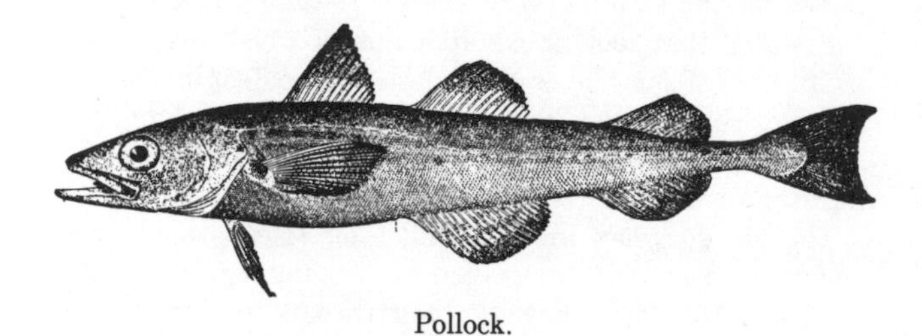

Pollock.

Flounder.

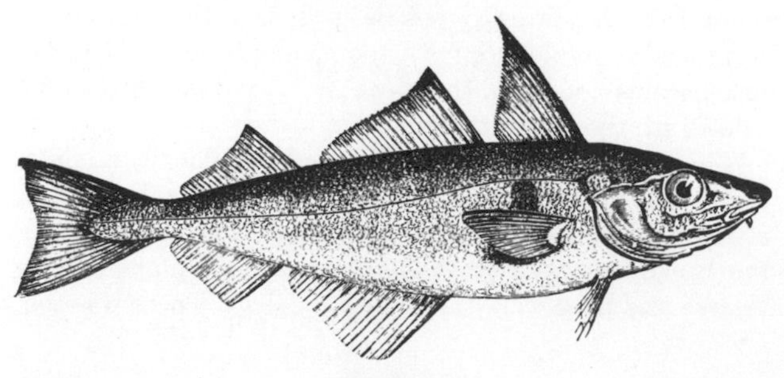

Haddock.

Halibut (discussed in the next section of this chapter) are the largest and most valuable of the flounder.

Rockfish are so named because they are found among offshore rocks or on rocky bottoms. Some examples are: striped or black sea bass, rock cod and grouper. Rockfish are unusual fish in that they bear living young.

The last decade and a half has seen a growing shortage of the traditional bottomfish—cod and haddock. This is because the demand for them has increased much faster than the supply. America's cod consumption rises, on the average, eight percent a year, whereas world cod landings only rise one percent a year. The obvious result of this situation is increased prices. Cod block prices have tripled since 1969. Biologists feel that there won't be any big increases in cod landings in the future, at least in the North Atlantic, where most cod are taken. There just aren't any more cod available.

In order to alleviate this situation fishermen are turning toward other groundfish, particularly pollock and whiting. New fishing areas like the Bering Sea off of Alaska are being developed. Japan has exploited this fishery much more than the United States has. In 1972 Japan landed 2.6 million tons of Alaska pollock, more tonnage than the *total* American catch that year of *all* fish.

The switch to other species of groundfish is progressing very slowly in North America, for Americans and Canadians are notoriously fussy about what they eat. It takes many years for a new type of fish to be accepted. Sometimes they have to be marketed under different names if they are to sell at all. For instance, prejudice against eating shark was partially overcome by calling it grayfish or seabass. Sablefish—a very tasty and nutritious sea creature that is excellent smoked—has to be marketed under the name black cod.

Only about twenty species of groundfish have any commercial value in American and Canadian markets. It seems inevitable, though, that we will eventually have to learn to eat new species of fish.

Halibut

This most valuable of the flatfish likes deep and cold water, 37 to 46 degrees Fahrenheit. It is found in the North Atlantic and North Pacific. It feeds on most other groundfish as well as on crab and clams. A mature male might weigh 50 to 100 pounds, but a mature female can weigh more than 500 pounds.

The word halibut comes from the Middle English words *hali* (meaning "holy") and *butt* (meaning "flounder"). The fish was called the holy flounder because it was eaten on holidays.

Halibut have been taken commercially in the North Atlantic for hundreds of years. The commercial halibut fishery in the Pacific began in 1888 in Washington waters. The fishermen learned much from the Makah Indians of Neah Bay, Washington. The Indians made hooks of wood and bone and lines of kelp or animal sinews. The hooks were weighted down with stones and the other ends of the lines were held up by buoys made of air-filled bladders. A two-man canoe handled ten or fifteen of these

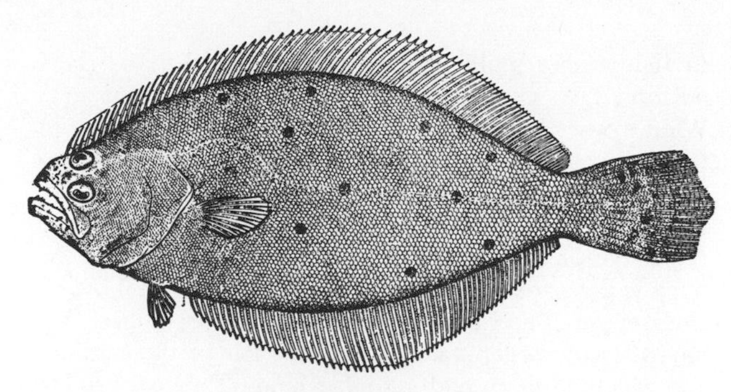

Halibut.

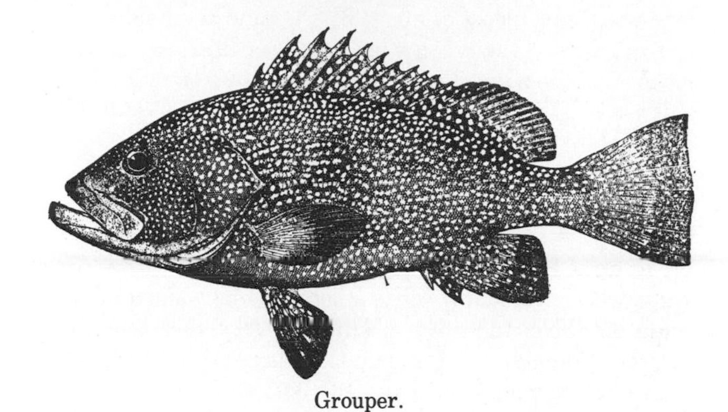

Grouper.

floating lines. Movement of a buoy indicated a strike had been made. In addition to being accomplished halibut fishermen, the Makahs regularly harpooned and beached large whales from their dugout canoes.

The halibut grounds that had been more than ample for the Makahs were soon overfished and depleted by the white men. After this happened the white fishermen simply moved north, depleting as they went. By 1911 they were working the waters off of Southeastern Alaska. By 1915 they had reached the top of the Gulf of Alaska. After this, they turned west and fished the deep ocean areas.

Pacific halibut landings hit an all-time low in 1931 due to gross overfishing. Strict conservation measures were set down to protect the vanishing fishery. By 1962 the conservation measures seemed to be paying off, for a record 75 million pound catch was taken that year.

Over the last decade the annual catch has declined steadily. This is not due just to depletion. It is in part due to the rigid restrictions placed on the fishery by the International Pacific Halibut Commission. It is also due to the shrinking size of the West Coast halibut fleet. Many fishermen are leaving for the more lucrative herring and salmon fisheries.

In the old days halibut were caught on setlines that were hauled into dories. A dory crew got twenty-five cents per fish, no matter what size it was. Each dory fished for itself. There were some grand fights when one crew thought that another had stolen its fish from the mother ship's hold.

If a dory crew was lucky it would be put down from the mother ship right in the middle of a chicken patch—an area of shallow water where chickens, or small halibut, were likely to be found. If it was not so lucky it would have to fish over a gully—a deep underwater canyon where nothing but whales (large halibut) were to be found. Whales were harder to haul and filled up the boat a lot faster than chickens. Since a chicken halibut was worth as much as a whale, dory crews hooking whales made a lot less money than those working a chicken patch.

One old-timer tells the story of how his captain always put the dory with the captain's relatives in it right over a chicken patch and put the other dories over gullies. One day, the storyteller was fishing over one of these gullies and was pulling up nothing but whales. They were so big and heavy that his hands were soon ripped and bloody and resembled raw pieces of meat. By the time he'd hauled only forty-five of the whales his boat was filled to the oarlocks. Then he noticed he'd drifted right into the middle of a dense king-salmon run. One by one he threw each of the whales he'd just caught overboard. Then he used everything in the boat that would float to buoy up his setline as close to the surface as he could, for that was where the king salmon swam. He even ripped out the benches and broke up a few of his oars and used them as floats, for kings were worth fifty cents apiece and he wanted all he could get. Soon there was a king on every hook near the surface. He made good money that day and it was easier than hauling whales up from the bottom of a gully. The skipper of course was mad as hell because his boat was practically torn to pieces, and fired him when they got to port.

Today, the Pacific halibut fleet is made up of aging men in aging boats. A 1963 study revealed that the boats averaged thirty years old and the crews were considerably older than those in other fisheries. A quarter of the men were over sixty. Halibuting is hard work—perhaps the hardest type of fishing. The boats sail far from shore on some of the roughest bodies of water in the world, such as the Gulf of Alaska and the Bering Sea, and they sail in winter when devastating storms are common. Perhaps this is why men, particularly the young, leave for easier and more lucrative fisheries. The crews are predominantly Scandinavian. The newcomers that do enter the fishery often do so because they have relatives already fishing in the fleet.

Sea Bass

Sea bass are, like many other fish, very sensitive to water temperature. Off North Carolina, for instance, black sea bass are found close to shore in summer, in shoal waters between 55 and 60 degrees. When winter comes and inshore water temperature drops, they head offshore and pick up the temperate waters of the Gulf Stream, a current that originates in the Gulf of Mexico and flows northeast to Northern Europe. North Carolina fishermen must follow the fish in their migrations, and have to travel up to four hours on winter mornings to reach the faraway fishing grounds. They often fish with traps. When a string of traps are set, they record their position by marking down the readings on their Loran set. The Loran readings will guide them back to the approximate location of the traps, close enough for them to sight the buoys the traps are attached to. Some of the North Carolina fishermen trap fish all year; others run party boats for sport fishermen in summer and trap during the off-season. On a good day, they will haul 600 pounds of fish into their boats. The fish are popular in East Coast restaurants, and bring about a dollar per pound at the buyer's.

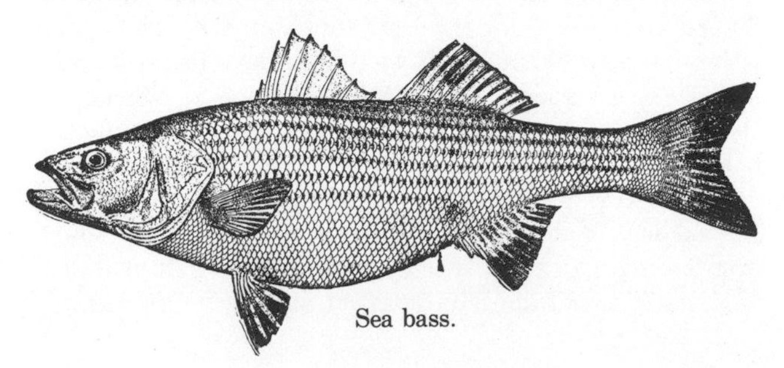

Sea bass.

Black sea bass are found near schools of small bait fish, upon which they feed. If the school is a large one, they are often too busy eating to notice any of the menhaden-baited traps fishermen have put in the area. If the school is small, they will usually smell the bait in the traps, and more will enter them.

Sablefish

Sablefish, known more commonly as "black cod," are very palatable and nutritious fish but until recently were regarded as next to worthless. The fish are found off the West Coast, at depths of 200 to 400 fathoms, deeper than most other commercially caught fish. They are not true cod, and they are not always black. Their backs are usually green to dark gray in color, and they average eight to ten pounds in weight. They are an oily fish, and are especially good smoked. When smoked, they are sometimes marketed as finnan haddie, although true finnan haddie is made from haddock.

Sablefish are taken in trawl nets and in setlines, and more recently in traps. In the past, they were usually taken as an incidental catch by boats mainly interested in halibut. Today, more and more boats are making sablefish their primary catch.

The new method of using traps to catch the fish produces fish of higher quality than those caught with trawl nets or setlines. Those caught on setlines were often partially eaten by the time the lines were hauled to the surface, and those caught in trawls tended to be crushed, scarred by the webbing, or pierced by the spines of rockfish that had also been caught in the trawl.

Several years ago, crab fishermen noticed that their traps often contained a number of sablefish. The Bureau of Commercial Fisheries became interested in this, and conducted experiments with modified Alaskan king crab pots, and cylindrical tunnel traps. The traps were connected to setlines and laid on the ocean's bottom. The experimenters eventually developed a collapsible wire mesh pot that weighed about 130 pounds. King crab pots weigh 600 to 800 pounds, and are more cumbersome and dangerous to use than the collapsible pots. The collapsible type can also be stored more compactly, and don't make a vessel so topheavy. The collapsible pots are the type most used today for sablefish.

Tuna, Menhaden and Other Fish

Tuna

A disease called "tuna fever" strikes Northern California every year about June. It begins with the landing of the first albacore tuna. The sickness is pretty well limited to troller captains of boats forty feet and over. The symptoms are severe anxiety and fantasies of dollar bills leaping aboard the boat.

The troller captains have to make a tough decision: should they continue catching salmon, or should they change all their gear and gamble on the less dependable albacore? A good albacore season can mean a fat bankbook and a leisurely vacation, but a bad season can mean going in debt and perhaps losing the boat.

It is common for a boat to fish for weeks catching barely an albacore, and then blunder into a school and land several tons in a couple of hours. This is why fishing albacore is such a gamble. The schools are few and far between, and if a skipper doesn't happen to find enough of them, he can go deep in the hole on his boat payments.

Albacore are pretty fish. Their backs are steely blue, shading into silver toward the belly. They are dappled with gold, bronze and green while alive, although their color fades rapidly after death. They are the most valuable of the tuna, and the only type whose flesh can be labeled "white meat."

Albacore Migration Paths

Albacore travel from Japanese waters to North America's West Coast and then back again each year. They appear off Japan from April through July and are taken there by the Japanese live-bait fleet. They arrive off the North American coast sometime between June and August and stay until fall, when they again head east. From October through March, they are found in the west-central region of the North Pacific Ocean and are taken by the Japanese longline fleet.

Albacore like water that is no warmer than 68 degrees and no colder than 60 degrees Fahrenheit. Ninety-five percent of the albacore catch is landed in waters within this temperature range. Their migration patterns are also affected by water temperature. During a year when ocean temperatures are low, the fish will stay further south, where the warmer water is. They will appear in North American waters as far south as Baja, California. As the season progresses and the water warms up, they will work their way northward, perhaps as far as Oregon, if water

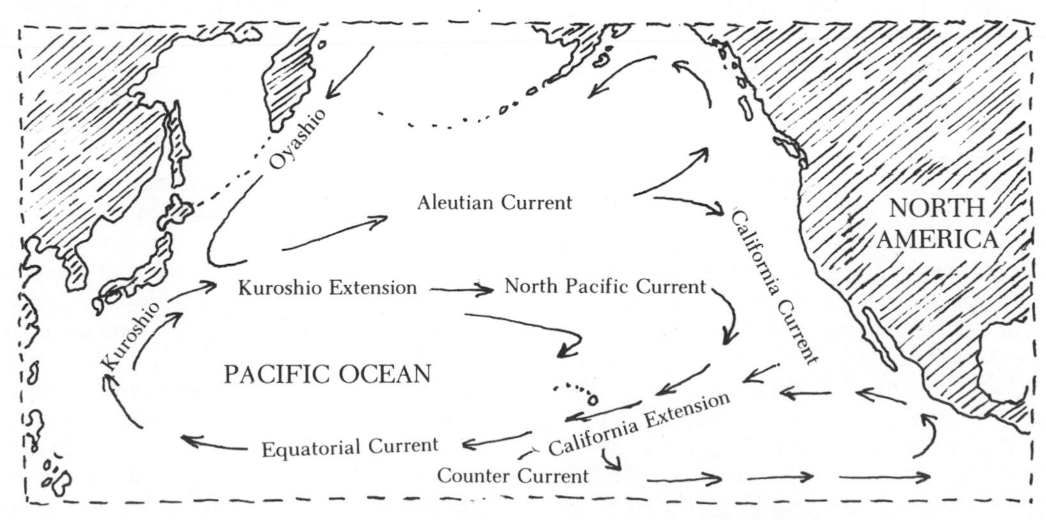

Current systems of the Pacific Ocean. *(John Douglas Moran)*

Fishing for tuna on a bait boat. *(National Oceanic and Atmospheric Administration)*

temperatures are low, or as far as Kodiak Island, Alaska, if the year has been unusually warm.

If there is a late spring and ocean temperatures remain cold for a longer time than normal, albacore might not be sighted off the West Coast until August. In years with an early spring they can be sighted in June.

Pacific albacore seem to move in patterns similar to some of the ocean's current systems. Eastward-moving currents in the Pacific split into two as they approach North America, one branch swinging north to form the Alaska Gyre and the other swinging south to form the cool, sluggish California Current. The albacore's paths of migration follow the eastward-moving currents and then also split into two branches, although somewhat to the southeast of the area where the currents split. The first albacore to reach North America take the lower branch, following the California Current and appearing off of Southern California or Baja, California. Later in the season the waters warm up and albacore heading toward North America take the upper branch, appearing off of Oregon or Washington. As the season progresses the fish work their way north along the coast. They are fish of the open sea and don't like to approach shore too closely. American fishermen rarely catch them within 30 miles from shore and often take them as far away as 300 miles.

Albacore are especially fond of squid, but they also feed on small rockfish, sauries and anchovies. Albacore average ten to twenty pounds in weight and generally travel in schools, although their schools are more spread out than those of other species of tuna, such as yellowfin. The largest known albacore weighed sixty-six pounds.

In the 1860s albacore were considered trash fish and were thrown overboard when accidentally caught. It wasn't until 1885 that it was discovered how well albacore could be preserved by canning. It took another twenty years, though, before the fishery started growing significantly. Around the turn of the century, albacore were taken as incidental catch on boats fishing for white seabass and bonito. Soon after this, boats began going out exclusively for albacore.

In the 1920s trolling became an increasingly popular method of fishing and other species of tuna were sought out as well as albacore. Many tuna clippers of up to 130-foot lengths were built during this period. The tuna clippers employed a live-bait pole and line method that will be discussed below. This method was learned from the

Unloading tuna at the cannery. *(Port of Los Angeles)*

Japanese and was the predominant one in the United States tuna fleet for many years. Today, most tuna are caught in purse seines—huge nets sometimes 500 fathoms (3000 feet) in length.

Tuna Boats

There are three types of tuna boats: clippers (also called bait boats because of the large tanks of bait they carry), jig boats, that use colorful lures instead of bait to attract the tuna, and seiners, that net the fish instead of catching them on hooks. The bait boats are fairly large—many are 130 feet or more in length. They hail from San Diego and San Pedro, California and are manned largely by first and second generation Portuguese and Yugoslav fishermen.

The bait boats catch three types of tuna: albacore, skipjack and yellowfin. When a bait boat locates a school of tuna, the live bait is taken from the tanks on deck and thrown to the school. This operation is called chumming. When a school is chummed, the fish will thrash and leap furiously as they gobble down all the food they can get to. Mounted on the stern and sides of the vessel are racks upon which the fishermen stand, holding bamboo poles attached to short lines, at the ends of which are barbless hooks. They dip the hooks into the water (as far from the ship as possible, for tuna are shy about coming too close to a boat) and the frenzied fish, ready to attack anything that is moving, strike the hooks and are hoisted aboard. The hooks, because they are barbless, can be quickly removed from the fish and thrown back into the water. If the fish are hitting well, a twelve-man crew can pull aboard twenty tons of tuna in an hour. The stern of the boat is trimmed down so that the fishing racks the men are standing on are only inches above the water, making it a lot easier to swing the fish aboard. Sometimes schools of fish remain "cold." They rise to within a few fathoms of the surface but won't eat the live bait fish that have been chummed into the water. In this situation, a handline baited with a live fish is lowered into the middle of the school to entice it to start feeding.

Jig boats are considerably smaller than the clippers (bait boats), for they need no heavy bait tank or large crew in order to fish. Many of the boats are only forty or forty-five feet long. They are usually fished by one or two men. Instead of bait they use "jigs"—lures festooned with feathers or strips of bright plastic that look like squid to the fish.

Tuna strike. *(National Oceanic and Atmospheric Administration)*

The boats speed along at up to eight knots and the jigs and hooks trail behind, attached to the boat with nylon lines. Up to fourteen hooks and lines dangle from the boat. When the fish strike they are hauled aboard by hand or by using a tuna puller—a small power gurdy that quickly reels the lines in. Jig boats are generally used for catching albacore only. Skipjack and yellowfin form denser schools than albacore and are caught more efficiently by seiners or bait boats with their larger crews and bigger holds. The huge seiners, on the other hand, have too-high operating costs to fish for the spread-out albacore. The seiners need to take thousands of fish in each set of their purse seines in order to show a profit.

As mentioned above, albacore boats sometimes troll for days or weeks without a strike. Then, all of a sudden, all fourteen lines will be straining with fish and hundreds more will be jumping madly for a chance to get at the jigs. When this happens, the number of fish caught is limited only by how fast they can be pulled aboard. Tuna are totally insane when they are in a feeding frenzy. They will bite a man's fingers off if he is foolish enough to dangle them over the side. They will slap the sea into a foamy lather with their jumps and gyrations. Experienced fishermen sometimes chum the school to prolong its feeding frenzy. Then, as suddenly as the fish appeared, they'll dive, and the sea will once again become calm. Boats might fish the same area for the rest of the day without a strike. It is not known what makes the fish abruptly decide to dive.

Advice for Fishermen

Old-timers offer many pieces of advice to the young albacore fishermen. Some of them are:

1) Wash the blood out of the fish's mouth as soon as you can. Blood in the mouth accelerates spoilage, discolors the flesh and gives the fish buyers a good excuse to cut the price they will pay you.

2) Don't let seaweed build up on your lines. An albacore jig is supposed to look like a squid. Fish aren't terribly smart, but they do know the difference between a glob of seaweed and a squid.

3) Start early. As in many other fisheries, the time shortly after dawn is often the best fishing time of the day.

4) Install refrigeration as soon as you can afford it. It seems it is always on the day your ice is used up that you stumble upon the largest school of tuna ever encountered by modern man and can take none of it, for you have no way of keeping it cool on the run back to port.

5) Never trust a fish buyer. If he smiles at you, be especially careful.

6) Understand that when you are catching tons upon tons of fish, so is everybody else, which means that fish prices will be low. It is when fish are rare that you will get good prices for them.

Superseiners

Tuna are big business. In 1974, 272 million pounds of yellowfin and 190 million pounds of skipjack tuna were landed by the United States and Canada. Some of the vessels that caught these fish cost over five million dollars to build.

Until the middle 1960s most skipjack and yellowfin were landed by live bait boats fishing with poles and lines. Two inventions drastically changed the situation, however. One was the development of the nylon purse-seine

Superseiner. *(Port of Los Angeles)*

net and the other was the invention of the power block for hauling the seine.

The gigantic purse seines that are used to encircle large schools of fish used to be made out of cotton and wore out in about a season. It was just too expensive to buy a new seine each year. Nylon purse seines last up to four years, however, and are thus economically feasible to use.

Purse seines used to be hauled by hand, and required large crews to do the backbreaking, rigorous work. The large crews needed large boats, and the whole operation was quite expensive—too expensive for the tuna industry.

In the middle 1950s Mario Puretic, a California fisherman of Yugoslavian descent, invented a power block that could haul purse seines quickly and safely, using much smaller crews than were previously used. A few years later the tuna industry discovered that they could now catch more tuna per dollar of expenditure with purse seines than with any other method. Scores of bait boats were converted to seiners and in 1962, the *Hornet*, the first of a long line of "superseiners," was built.

Yellowfin tuna often travel with schools of porpoises. Fishermen use this fact to locate and catch the yellowfin. Tuna are hard to spot from the boat because they travel under the surface. Porpoises on the other hand travel on or near the surface, for they are mammals and must surface in order to breathe. The Pacific spotted dolphin, or spotter, is the type most often found traveling with yellowfin. No one knows why spotters and tuna like to stay together. Perhaps it is because the spotters protect the yellowfin from shark. Perhaps it is because of similar eating habits (yellowfin diets do resemble those of spotters). Or perhaps it is because of a need for companionship. Yellowfin have also been known to congregate around logs or floating clumps of seaweed.

Tuna seiners are equipped with high-powered binoculars on the bridge and in the crow's nest. The lookouts search not just for porpoises but for several species of birds as well. Terns, boobies and frigate birds often hover above schools of fish that the tuna like to feed on. The lookouts also keep their eyes open for "jumpers," "shine" and "black spots." Jumpers are tuna that break the surface as they feed. Shine is a flash of light reflected from fish below the surface. A black spot is a dense school of fish that is barely visible under the water. If any one of these sightings is made, the boat prepares to set its purse seine (see the chapter entitled "Fishing Gear").

Before the seine is set, the school must be herded into a tight bunch so that they can be encircled by the net. To do this a "pongo"—a high-powered skiff fast enough to outrun the speedy yellowfin and porpoise—is lowered from the boat. Large seiners usually carry two and sometimes three skiffs aboard. They are about sixteen feet in length and are powered by big 85- or 105-horsepower outboards. Skiff drivers wear crash helmets and strap themselves into their seats, for these boats can achieve speeds of up to forty knots. The chase is directed by radio from the main boat by the fishing captain, who may or may not be the ship's captain.

When the porpoises sense the skiff approaching, they run and the tuna follow them. The pongo roars in pursuit, trying to head them off and herd them into a tight bunch so that they may be easily netted. The herding operation resembles a cowboy working cattle.

Sometimes the porpoises won't allow themselves to be herded but keep running all out. When this happens the pongos try to at least get them to run in a tight group, hoping the tuna will do likewise. They also try to turn the porpoises in a downwind direction. Tuna closely follow porpoises running with the wind, but tuna will outdistance porpoises running against the wind and will be harder to capture.

The ship and another skiff, the seine skiff, are maneuvered so that the tuna will swim between them. Strung between them is the purse seine. After the fish swim into it, the ship and seine skiff drag the ends of the net around so that the fish are surrounded by it. The fishing captain waits until the porpoises that have been surrounded calm down. Then he orders that the purse lines of the net be slowly hauled in so as not to startle the fish. Hauling the purse lines closes the bottom of the net and prevents the escape of anything contained in it. If the fish don't calm down but stay restless and dive often, the purse lines are hauled as soon as possible before the school dives under the net and escapes. If even a few porpoises get out, the whole school of tuna is likely to follow.

Backing Down

After the seine has been pursed, the captain performs the "backing down" operation. Its function is to give the porpoises a chance to escape while keeping the tuna imprisoned. The porpoises usually form a very tight group, as far away from the boat as possible. A large bag or balloon is put into the net near the boat. This balloon is supposed to attract the tuna the same way they are attracted to floating logs or buoys, and keep them away from the porpoises. The skipper then backs the ship up, sinking the far end of the net and allowing the porpoises to swim to freedom.

There are differing reports on how effective "backing down" is in allowing porpoises to escape. Some drown or get severely injured by entangling themselves in the webbing of the net as they try to swim across. Some refuse to swim across and get hauled up with the net. Randall Reeves claims, in an article in *The Nation* (24 May 1975), that in a single set as many as a thousand porpoises get trapped in the net and drown.

Modern purse seiners use nets equipped with the Medina Safety Panel, a strip of fine mesh webbing in the back-down area, that supposedly prevents porpoises from entangling themselves. In the old nets porpoises often got their snouts and appendages caught in the larger mesh webbing.

Tunamen don't want to kill porpoises if they don't have to. They realize that their livelihood depends on cooperation with them. Besides, most fishermen like the porpoises. Neither the back-down procedure nor the Medina Safety Panel were invented by conservationists. Both were invented by tuna skippers. These inventions seem to have worked to some extent, for the estimated spotted porpoise kill declined from over a quarter-million in 1972 to less than 100,000 in 1974. The reason for the low kill rate, though, could have been that most of the porpoises in the fishing areas have already been wiped out. Some scientists estimate that the present population is only one-third of what it was several years ago.

There are indications that porpoises might be learning to evade capture. More and more often, pongos have to chase the schools for unusually long periods of time before they are able to herd them into a tight bunch. Increasing numbers of sets are total failures, catching neither porpoises nor tuna. And then there are the "untouchables"—dolphins that consistently manage to outrun, outmaneuver and outsmart the pongos.

If porpoises do learn to avoid capture it would certainly hurt the tuna industry, for most yellowfin caught in the Eastern Pacific were traveling with dolphins when they were netted. As the 1976 season got underway, tunamen were warned to decrease kill rates by 30 percent or be forbidden to take any dolphins in their nets at all. And in May 1976, a United States district judge ruled that tuna fishermen could no longer ensnare porpoises in their nets along with the tuna, for they had failed to meet the 30 percent kill-rate reduction. The general manager of the American Tunaboat Association responded to the ruling by saying that the ban on porpoises could mean an end to the industry. It will certainly put many Southern California tunamen out of jobs unless alternate ways to catch yellowfin are developed. One fisherman interviewed expressed views common to many of his shipmates:

"The environmentalists think that we are devils. We don't want to hurt dolphins, because we need them to find tuna. We save them when we can. What the environmentalists don't realize is that now, some other country will catch the yellowfin we are not allowed to catch; but they won't be so careful about how many porpoises they kill. More porpoises than ever will die because of the ban."

Menhaden

The fish caught in the most abundant numbers in North American waters is a fish that few people have ever heard of—the menhaden. The word menhaden comes from the Algonquin Indian word meaning "they fertilize." The Indians used to plant the fish with their corn to insure a good crop. The menhaden is known by many other names. Some of them are: pogy, whitefish, hardhead, bony fish, alewife, shiner, green tail, bugfish, fat back, yellow tail, mossbunker and just plain bunker.

The menhaden's lack of recognition is perhaps due to the fact that it never appears in supermarkets or fish stores, for it is considered inedible. It is used instead for "reduction"—it is crushed, ground up, and made into fish oil, fish meal and fertilizer. The fish meal is used as a high-protein additive in poultry and swine food, and the fish oil is used in margarine, paint, ink, special coatings and in many other applications.

The fish are caught along the Atlantic Coast and in the Gulf of Mexico using purse seines. The seiners often work in conjunction with small airplanes, whose job it is to spot the pods, or schools, and guide the purse boats to them. Sometimes the planes remain above a school while the net is being set, and radio instructions down to the boats. A big pod of pogies often contains more than a million fish.

The average pogy weighs less than a pound, and is about a foot long. After the fish are caught in the purse seine, they are pumped into the holds of the "steamer" using large centrifugal pumps. Almost half of all menhaden caught in North American waters are landed by Seacoast

Hauling the menhaden purse seine. *(National Oceanic and Atmospheric Administration)*

Pumping menhaden into the hold. *(Marco)*

Products of Port Monmouth, New Jersey, which makes them the largest purse seining operation on the North American continent.

The year 1973 was an exceptionally good year for pogy fishermen. The fish appeared early in the year, stayed late, and by the time the season was over, almost one billion pounds had been landed. A large catch can often drive prices down disastrously low, but just the opposite occurred in 1973. The world soybean harvest was bad, and the huge Peruvian anchovy fishery was in trouble. Soybeans and anchovies are used for many of the same applications as menhaden, and their scarcity drove menhaden prices up to record heights.

Menhaden boats at times come into near-violent conflict with sports fishermen. Sportsmen do not want the pogy schools to be destroyed, for they claim that pogies are a favorite food of the bluefish, one of the East Coast's most popular game fish. They say that bluefish become scarce in areas that have been heavily fished for menhaden. A few years ago, incensed sportsmen in Long Island Sound surrounded a menhaden steamer with their boats, and wouldn't permit it to leave the area, or even to retrieve its catcher boats. The tide began carrying the whole assemblage close to some dangerous shoal waters, and the Coast Guard had to be fetched to break things up before the steamer ran aground.

Sportsmen claim that the seiners and their large nets endanger the small recreational craft, and that the seiner's pump emissions cause oil slicks and make messes of nearby beaches. In addition, pogies have a rather strong smell, and wastes dumped by the boats often carry the smell to the shore.

Although the huge purse seines capture many millions of pogies each year, the fishery shows no signs of depletion. The menhaden catch in 1876 was about half a billion pounds, and the annual menhaden catch today still averages about half a billion pounds. Thus, claims from the sports fishermen that menhaden stocks are being depleted seem to be unfounded. Seacoast Products says that each year about one-third of the pogies in each of the heavily fished areas are landed in purse seines, another third are eaten by natural predators (like the bluefish), and another third escape and are able to spawn. Experts feel that enough pogies escape to replenish the stocks.

A few summers ago, thousands of dead pogies washed up onto Milford, Connecticut beaches. The sportsmen blamed the purse seiners, but when the fish were examined by a laboratory, they were found to have died from a natural disease. Some scientists feel that this disease is one of nature's ways of preventing menhaden from becoming too numerous.

Menhaden spawn in the spring in northern waters, sometimes as far north as the Gulf of Maine, although usually somewhat south of Cape Cod. The eggs hatch at sea during summer. The newborn fish spend the summer in coastal estuaries, moving south when temperatures

Menhaden fishermen and seine skiffs, Southport, North Carolina. *(National Oceanic and Atmospheric Administration.)*

drop in autumn. Plankton, a major food source of the pogy, don't reproduce fast enough in cold water to support the huge pogy schools, and the fish are forced to migrate to warmer waters. The next summer, the pogies again move north. The older a pogy is, the farther north he travels in summer. Two-year-olds get as far as New York harbor. Three- to five-year-olds reach Cape Cod and points north.

Swordfish

The lookout is in the crow's nest, watching for a fin much like a shark's, but held steadier and more erect. He spots one, and the boat takes off in pursuit. The harpoon man grabs his long spear and leaps out on the pulpit, a small platform built on the boat's bowsprit. The boat overtakes the swordfish, and the harpooner heaves his lance with all his strength. His aim is good, and the harpoon hits home. The fish takes off running, but the head of the harpoon has been driven deep into its flesh. A strong line runs from the harpoon's head to a buoy, which is thrown overboard when the swordfish is harpooned. A dory is lowered, and the doryman picks up the buoy. The fish fights gamely, but he has to pull a heavy dory, and his energy soon runs out. When he tires, he is reeled in and killed.

This is how swordfish were caught seventy-five years ago, and this is how they are caught today. The success of a trip still depends on the keen aim and strong throwing arm of the harpooner, and on the teamwork of the crew.

The harpoon has not changed much in the last century. At its end is a dart called a "lily." After a strike is made, the rest of the harpoon comes loose and is hauled aboard the boat. The lily however is curved, and when the line attached to it is pulled taut, the lily twists sideways, making it impossible for the swordfish to pull free of it.

Modern swordfishing operations often use a spotter plane to help locate the fish. The single-engine plane flies in circles around the boat, radioing locations of swordfish down to the captain. The harpooner also listens in, so that he may know where to look for the fish. The pilot, often an independent operator, gets fifty to seventy-five dollars per fish. The radioed instructions go something like this:

"Captain Harry, there's a big one at three o'clock, oh, about twenty boat lengths away. Turn a bit more to starboard. That's it, you're heading straight for him. He's moving south. Turn a few more points starboard. Good! Distance about eight boat lengths. Six lengths. He's at eleven o'clock now. Swing back to port a bit. Larry, get your spear ready. You'll reach him in a few seconds. Good shot!

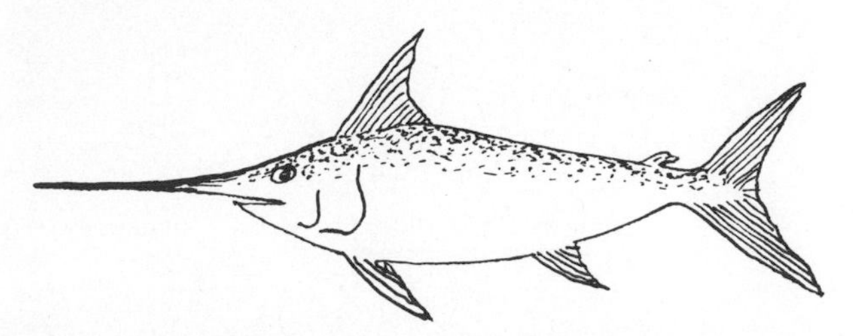

Swordfish. *(John Douglas Moran)*

Right on the money. Put a dory on that baby and come about quick. There's another one directly astern, not half a mile away. . ."

In the old days, there were no spotter planes, and as many as five men climbed the mast as lookouts. The first thing seen were the fish's fins. If the fish was a few feet below the surface, all that might be seen was a reddish blue blur under the water. The boat had to get pretty close to tell if it was a swordfish or shark.

The most dangerous part of swordfishing comes when the doryman is hauling in the exhausted fish. Sometimes a swordfish will get a new burst of energy and charge the dory. Its hard and sharp sword can go through the side of a wooden dory as if it were cardboard. And if the doryman happens to be in the path of the sword. . .

Swordfish are almost unique in that there is only one species of the fish—the broadbill swordfish—in all the world's oceans. The official name for it is *Xiphias gladius*. *Xiphias* is a Greek word meaning "sword-shaped." *Gladius* is the Latin name for the short sword that the Roman infantrymen used to carry.

The sword is an extension of the fish's upper jaw. The swordfish uses this sword to slash its way through a school of smaller fish like herring or mackerel. After it cuts its way through, it will dive down a few feet, turn, and swim back under the school, catching and eating the dead or maimed fish as they sink. The bones are eventually vomited back out.

There isn't much in the sea that the swordfish runs from. Killer whales, sperm whales and humans are its only enemies. This fearlessness works to the fisherman's advantage. He can often bring his boat to within a few feet of the swordfish, making him easy prey for the harpoon.

Most of the commercial harpoon boats work out of Atlantic ports. There is a small fleet, though, on the West Coast, based in the Southern California ports of San Diego and San Pedro.

Swordfish are also taken with twenty-mile longlines. Baited hooks are fastened to the line about every fifty feet. The line is buoyed up with floats to keep it near the surface. Every once in a while, a float is omitted, allowing that section of line to sink lower and catch the fish that happen to be swimming at a deeper depth than normal. In most other types of North American longlining (also called setlining), the line is allowed to sink right to the ocean floor, for that is where the groundfish (cod, halibut, haddock, etc.) are.

Swordfish feed by night, and so the line is set in the evening and left until the next day. If all goes smoothly, the lines are hauled up by noon, and the crew gets a chance to rest for a few hours. If there are delays, as there often are, hauling continues until well into the afternoon or evening, after which the thousands of hooks must be immediately rebaited and the line returned to the water.

A few years ago, the Food and Drug Administration discovered that some swordfish contained amounts of mercury that were possibly dangerous, and the fishery was temporarily closed. In 1972 the FDA announced that they could only prohibit the selling of swordfish if interstate commerce were involved. The FDA could do nothing if the fish never left the state they were originally sold in. This decision allowed the fishery to open once again.

Sharks

The only time shark meat has been sold openly as such and still sold in large quantities was in World War II. Due to meat rationing, Americans and Canadians were desperate for any type of meat or seafood that they could get their hands on. In other years, it was sold under more marketable names such as grayfish or sea bass. Many North Americans are repulsed by the idea of eating shark. One fisherman told me he would as soon eat his dog.

There are 250 to 300 species of shark, the biggest of which is the whale shark, often attaining lengths of forty-five feet or more. Sharks are tremendous nuisances to fishermen, for they often eat the catch before it can be hauled aboard. They do their worst damage in the Gulf of Mexico and in southern Atlantic waters. They will rip through nets laden with shrimp or fish in seconds, leaving large, gaping holes that take considerable time to mend. One Florida mackerel fisherman told about three particularly bad days when sharks tore ninety-one holes in his nets. It took two men thirty hours to fix the nets, and a lot of valuable fishing time was lost.

At the "Shark and Man" conference held recently in Florida, it was recommended that a larger commercial fishery for sharks be set up. This would reduce shark attacks on nets and on bathers as well. It could also be a quite profitable enterprise for a small fleet of boats. A seventy-eight foot vessel is now being outfitted for shark fishing in southern United States waters. It will lay thirty miles of line a day, and will have facilities for washing, bleeding, skinning and icing the catch on board. To prevent spoilage, these operations must be done within a day after the shark is landed.

Every bit of a shark's body has a use. Fins are exported to Hong Kong, where they are used to make shark fin soup. The teeth are made into charms and necklaces. Places such as Ocean Leather in New Jersey buy the hides and make them into tough, pliable, scuff-resistant leather. The leather is used in expensive boots, purses and luggage. The meat is usually exported to South America and Europe, although some fast-food chains are experimenting with using it. The liver is a rich source of vitamin A. Shark's livers used to be one of North America's main sources of this vitamin before synthetic vitamin A was developed. The shark's cartilage and offal are ground up and made into fish meal, and its skeleton is made into glue and gelatin. A 500-pound shark will produce about 150 pounds of usable meat, 100 pounds of liver and 2½ pounds of dried fins.

Sardines

The Pacific sardine fishery used to be the largest fishery on the West Coast, but is now virtually dead. Fishermen claim that its demise was due to gross overfishing by the California and Pacific Northwest purse seine fleets. Sports fishermen agree, and warn that the same thing will happen in the anchovy fishery if strict controls are not imposed.

Some scientists say that the sardine was not fished out of existence, but simply left the area. There is evidence in fossils that the sardine migrates around the world in cycles of 500 to 1700 years. They only spend a small part of this time on the West Coast. When they leave, the anchovy takes their place. Thus, the coming and going of these fish is perfectly natural. Other theories speculate that it was the low water temperatures of the 1940s that chased them away. The low temperatures are believed to have had an adverse effect on the albacore fishery.

Before World War I, more money was made in the sardine fishery than in any other on the West Coast. Monterey, California was the center of activity. John Steinbeck's *Cannery Row* gave the town worldwide fame. Monterey's Cannery Row at one time had eighteen sardine processing plants, supplied by seventy-eight purse seiners. When the fishery fell apart in the 1940s and early '50s, many of the boats remained unused for years. They were too small to be good tuna clippers, and too large to fit into the fifty-eight foot Alaska limit boat class. Many eventually saw use in the Alaskan king crab fishery.

Sardines have been caught in other parts of the country, particularly in Maine, but their numbers never equalled those landed on the West Coast. Today, Maine is the only state that processes sardines for human consumption. Maine's industry is being hurt by imported Norwegian, Portuguese and French sardines. Between 1965 and 1970, foreign-produced sardines expanded their share of the total American consumer market from 35 to 65 percent. What the foreign companies gained, Maine lost. Maine is in the process of seeking market protection from the United States Tariff Commission against these imported products.

Mollusks and Crustaceans

Oysters

Oysters are usually found in quiet, shallow inlets. An average oyster lays up to half a billion eggs in a season, throwing them out in a milky spray. The larvae, called spat, are hatched ten hours after the eggs are fertilized and immediately begin to swim. They spend the first two weeks of their lives looking for rocks, bits of shell or other hard surfaces. Most of the spat are eaten by fish, snails, starfish and other marine animals before they find a place to attach to. When they find such places, they attach themselves and stay for the rest of their lives, which can be up to twenty years long. Tons of broken shell are dumped into commercial oyster beds each year to give the spat plenty of spaces on which to settle.

On each side of the spat's body are folds of tissue called mantles. The mantle excretes a limy substance that hardens to become the shell. If a parasite or grain of sand gets inside the shell and irritates the oyster's soft body, the mantles excrete lime over the sand grain and eventually a pearl is formed. The best pearl oysters are found in the tropics. Most of them are not edible like the oysters of the earth's temperate zones.

In shallow water, the mature oysters are harvested from the bottom with huge pincers, called tongs, that open and close like a pair of scissors. The tongs have basket-like ends that scoop the oysters from the bottom when they close. In deeper water, dredges—baskets made of iron and net—are dragged along the bottom.

Tonging for oysters.

Oystering Under Sail

The last commercial sailing fleet in North America is on Maryland's Chesapeake Bay. It harvests the bay's oyster beds both by tonging and by "drudgin'." The dredge boats are particularly sensitive to weather conditions, for if there is either too much wind or too little, the boats must stay in the harbor.

Fifty years ago, the bay was filled with schooners, sloops, bugeyes and skipjacks, and there were enough oysters for everyone. Today, catches have decreased drastically, and the skipjacks are the only sailing craft left.

Skipjacks have been built since the 1880s. The first ones were flat-bottomed, but they were later made with V-bottoms so that they could sail closer to the wind. They had centerboards that could be raised when they worked in shallow waters. To give them more ballast, a couple of tons of fieldstone were put on board. Some skippers used to use old tombstones. It's rumored that some of the older skipjacks still have tombstones lying under their bunks.

Skipjack dredging is a vanishing art. The boat's speed is critical. If it sails too fast, the dredge won't stay on the bottom, and if it goes too slow, the dredge might hang up on something. The only way to lower the boat's speed is by reefing its sail—reducing its surface area by letting down a bit. Chesapeake skipjacks have four reef bands on each sail, giving them four speeds.

The firmness of the bottom is tested with long poles or by feeling the tension in the dredge cables. Muddy bottoms won't produce oysters, for the mud smothers the spat. The captains look for hard bog bottoms. The biggest oysters are on the edge between the hard bog and the mud, because that is where the most food is.

Other Mollusks

Abalone are found on the West Coast's rocky shores and reefs, from the intertidal zone down to 500 feet. Unlike oysters, they have only one shell. It takes a steel bar and a fair amount of strength to pry one loose from a rock.

North America's only commercial abalone fishery is in California. Four to five million pounds are harvested annually, and it is illegal to export them to other states. Chinese immigrants started the fishery in the 1850s. They worked out of small boats, prying the mollusks loose with iron-tipped rods. Japanese entered the fishery after 1900, and were the first to dive for abalone.

A typical abalone boat is thirty to forty-five feet long, and carries a boat handler, a line tender and one or more divers. Abalone are found around kelp beds, for this is their favorite food. The diver must slash through the kelp to get at the mollusks, and it often rises to the surface and fouls the boat's rudder and propeller. The divers put their catch in mesh baskets attached to the boat with a line. When a basket is full, they tug on the line and the boat crew hauls it up.

After the baskets are unloaded, the abalone are left for several hours until their powerful muscles relax. They are then shucked from the shell, sliced into thin steaks and pounded with wooden mallets until they are tender.

The squid is similar to other mollusks in that its body is soft and boneless, but it differs from oysters and abalone because its shell, or pen, is on the inside of its body. It ranges in size from several inches to forty feet, and is closely related to the octopus. Squid are caught mostly for bait or reduction, although Italians and Chinese consider them delicacies.

Squid are usually caught at night in nets. The densely packed schools rise to the surface as it gets dark. They disturb the upper layers of zooplankton, causing them to glow. Schools of fish also make the zooplankton glow, but not as uniformly as do schools of squid. Also, schools of fish change direction more often than squid.

Squid.

The scallop has two shells like the oyster, but it does not remain in one spot for its whole life. It is able to swim by opening and closing its shells like a pair of flippers. It has been caught on the East Coast for many years, but is one of the West Coast's youngest fisheries. The Gulf of Alaska enjoyed a short scallop boom in 1968. There were 1.1 million pounds landed, which was a national record. By 1970, scallop landings had already begun to wane. Martha's Vineyard, an island off Cape Cod, Massachusetts, seems to be in the middle of a similar boom.

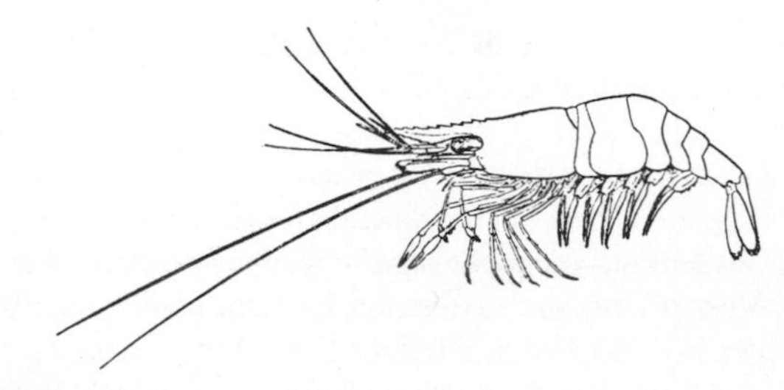

Shrimp

Shrimp, like crabs and lobsters, are crustaceans. They wear their skeletons on the outsides of their bodies. These exoskeletons form suits of armor around them, and act as protective devices. Their armor is not able to grow, however, and this is why crustaceans shed their shells from time to time and grow larger ones.

More money is made in the United States from shrimp than from any other sea creature. Although the yearly tonnage of shrimp is only a small fraction of the total of all fish caught off North America, the value of the shrimp catch is about one-quarter the value of the total catch. In 1960, for instance, shrimp landings made up only 8 percent of the total United States catch, but were worth about 106 million dollars, 24 percent of the total.

Shrimp are found off most of the North American coast. They are caught from Nova Scotia down to the Gulf of Mexico, and from Mexico up to Alaska. But the best shrimp fishery is in the Gulf of Mexico. Enormous profits have been made there since the discovery of pink shrimp in the early 1950s. Pink shrimp come out of their muddy hiding places at night only, and the old daytime shrimp boats dragging for whites didn't even know of their existence.

Shrimp go through several stages of development before reaching maturity. In their early, or larval stages, they resemble mullets or baby sailfish, and it takes an experienced biologist to identify them as shrimp.

Most shrimp spawn only once, and live to the age of one year. The female releases up to a million eggs when she spawns. The babies drift with the currents. If they don't get eaten by arrowworms, young fish or crabs, they eventually sink, hopefully in a shallow area with a muddy bottom.

Dragging for shrimp, Galveston, Texas. *(National Oceanic and Atmospheric Administration)*

Development of the Gulf Fishery

In the years before the First World War, Gulf of Mexico shrimp were caught in beach seines that were dragged over the mud of shallow bays. Some of these nets were quite small, and could be hauled by only two men, while others were half a mile long and needed twenty men to pull them. All were designed for work close to shore.

Between 1912 and 1915, Portuguese and Italian fishermen from Fernandina, Florida began experimenting with the otter trawl, a type of net that is dragged by a boat over the sea bottom, and can be fished in waters several hundred fathoms deep. The experiments were quite successful, and allowed the abundant offshore shrimp beds to be harvested. Boats were able to fish with smaller crews, too, for otter trawls can be handled by two or three men. In a few years, shrimping became immensely profitable. In 1889 less than ten million pounds of shrimp were landed. By 1930 the annual harvest had jumped to eighty-eight million pounds, and was still growing. Shrimping reached a peak in the 1940s, and a federal government study declared that no new grounds were likely to be found.

Then, in 1950, the Salvador brothers discovered a type of shrimp that fed at night, rather than during the day, as white shrimp do. They had been searching all day for new shrimp grounds off of Florida, in a place called the Dry Tortugas. It was growing dark, and they decided to lower their trawls one more time before going home. When they hauled the nets back up to the surface, they were filled with pink shrimp that had just come out of their burrows.

The Salvador brothers tried to keep their find secret, but the word got out and within a week, "pink gold fever" was sweeping through the Gulf ports, and 400 vessels were dragging the Dry Tortugas beds.

Today, the Gulf catch is comprised mostly of pinks and Golden Brazilian, or brown shrimp. Whites make up only a small percentage of the total landings, although they are worth more per pound than the pinks.

Life on a Shrimp Trawler

A typical shrimp trawler has a crew of three men: the captain, the rig man and the header. The rig man's job is to see that all the gear is in good repair. The header, the low

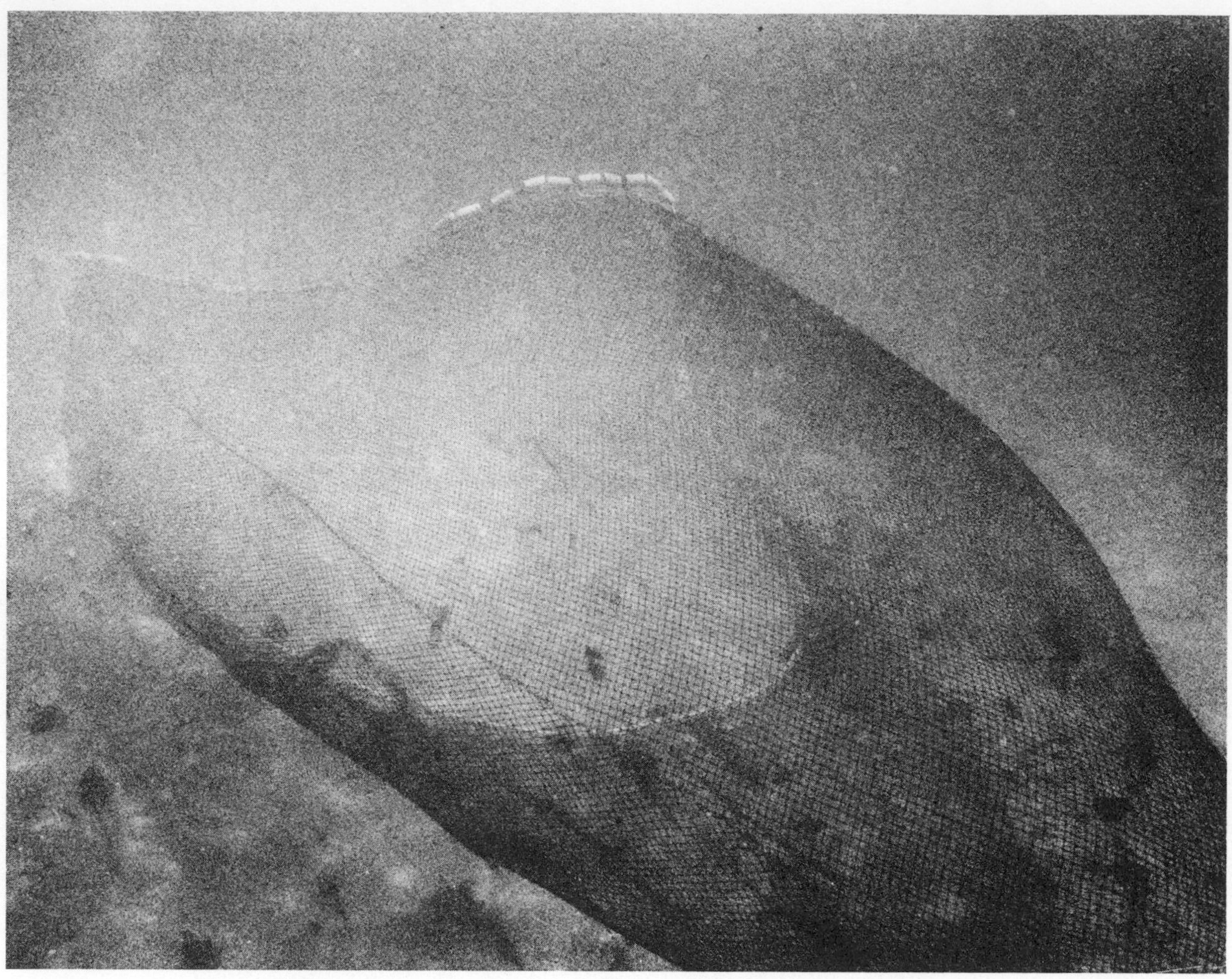

Shrimp trawl in operation. *(Seattle Marine and Fishing Supply Company)*

man on the totem pole, does most of the dirty work, like breaking the heads off shrimp and packing them carefully in ice.

When a boat arrives at the grounds, it throws a small try net over the side and drags it on the bottom for a few minutes. If it comes up with a fair number of shrimp in it, the trawl nets are lowered. These are similar to conventional otter trawls, but are much smaller. Bright colored flags are attached to the trailing ends of the trawls in order to keep sharks away, for a shark will tear a hole in the net the size of a Greyhound bus trying to get at the shrimp. Electric shock devices, sonic signals, bubble curtains and chemical dies are also used, although it is questionable if any of these devices do much good. Once a shark decides to feed on something, there is not much that can stop it.

The two trawls are attached by long ropes to two booms, called outriggers, that hang over each side of the boat. The trawls trail behind the boat, dragging along the bottom and scooping up shrimp, shark, eel, sculpin, hake and anything else that happens to be hanging around. The small try net also continues to drag behind the boat. It is hauled up every half-hour and checked. If there are only a few shrimp in it, the captain will order a course change. The trawls are hauled every four hours, unloaded on the deck, and immediately returned to the water, for a boat can't make any money if the nets are sitting on the deck.

Various things besides sharks will damage a trawl net. Sawfish will hole a net in seconds if caught in it. The trawl often catches and rips on rocks. In North Carolina fisheries, the drag boats must stay in water forty feet deep or less if they don't want to rip their nets on the rocks that lie at deeper depths. The shrimpers who fish out of Southport, North Carolina have to be especially careful, for there are several sunken Civil War vessels nearby that will play havoc with an otter trawl. The skippers in the area know their exact location, and are usually able to avoid them.

When a boat snags on a rock, the engine will start laboring noticeably and the skipper will have to throw the boat in reverse and back the net off the rock. Then he'll have to haul it, repair it on the spot, and return it to the water.

Boats dragging for cotton (pink shrimp) drag from dusk until dawn. In tropical areas like the Gulf of Mexico, this

Shrimp trawler with outriggers raised. *(National Oceanic and Atmospheric Administration)*

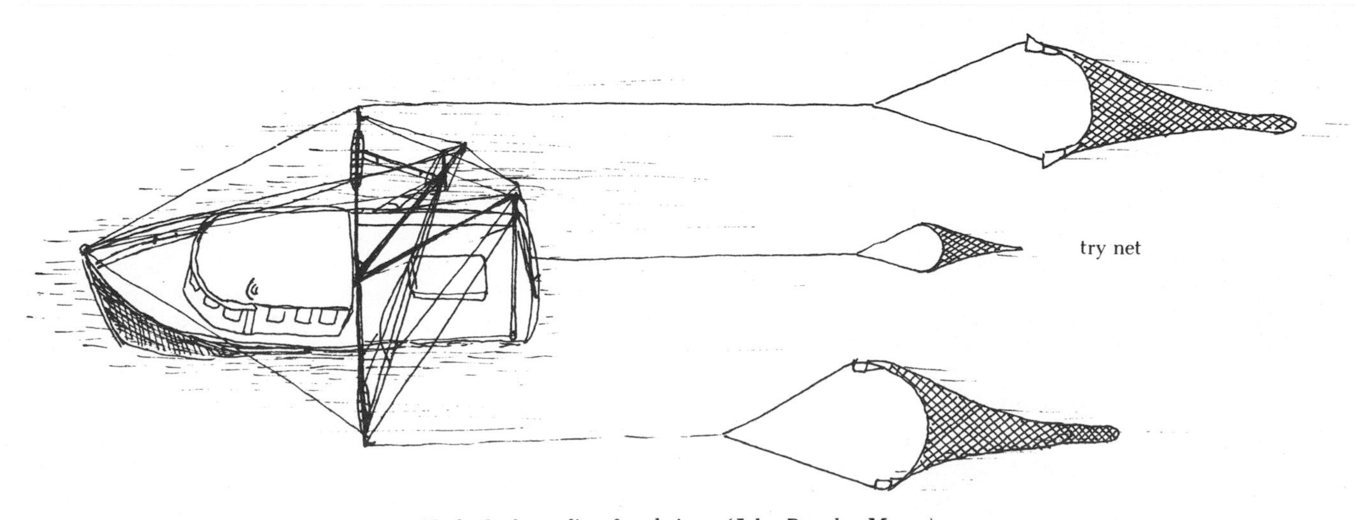

Method of trawling for shrimp. *(John Douglas Moran)*

means from about 6 P.M. to 6 A.M. The crew sleeps or repairs gear during the daylight hours. A trip lasts perhaps eight days. At the end of the trip, many boats return for only four hours, after which they immediately leave on another trip. The men from the packing company will meet the boat at the dock, suck up its catch in large hoses, refill the hold with ice, refuel the boat and replenish the supplies of food. Some fishermen say that the whole operation resembles a pit stop in a Daytona stock car race.

Sorting and Processing

Different boats haul their nets at different times, depending on the size of their nets and on the abundance of shrimp. Sometimes, nets are hauled every twenty minutes, and at other times, not for four hours. When the trawl is hauled, it is lifted clear of the water and swung over the deck. The bottom of the trawl is opened, and several hundred pounds of sea life tumble out. Perhaps a third of it is shrimp. The rest—shark, hake, flounder and many other types of sea life—are trash, and are immediately swept overboard, so that the fish have at least some chance to survive. It seems questionable, though, how many fish can survive being dragged along in a trawl net for four hours, squashed tight against several hundred other fish.

As the men sort out the shrimp from the trash, they must be careful of any red scorpion fish that were picked up in the trawl. Their needles are poisonous, and can inflict painful wounds upon those who unwarily pick them up.

After the trash fish are swept overboard through the scuppers, the shrimp are processed. The crew begin picking up the shrimp, one in each gloved hand, and behead them by putting the thumb on one side of the shrimp's head, the forefinger on the other side, and flicking. A shrimp's head is 40 percent of its weight, and contains its heart, stomach, and most of its internal organs. Thus, beheading a shrimp pretty much cleans it, and lengthens the time it will keep without spoiling. A good header can fill a sixty-pound basket with beheaded shrimp in half an hour. The baskets are then washed with hoses and lowered into the hold, where the shrimp are packed in bins with crushed ice. First, a layer of ice is sprinkled on the bottom of the bin, then a layer of shrimp, then another layer of ice, and so on. Care is taken to see that no two shrimp touch each other for they will quickly spoil if they do. Every shrimp must be surrounded by a layer of ice.

On boats equipped with refrigeration, the shrimp are frozen in chilled brine or in an air-blast freezer, and are stored in a refrigerated hold. In order for shrimp to retain their good flavor, they must be frozen or iced soon after they are caught.

Sometimes a boat will catch different species of shrimp at different times of the year. For instance, in the North Carolina shrimp fisheries, in areas such as Baldhead Island, Corncake Inlet, or Frying Pan Shoals, spotted shrimp are caught in April and May, and brown shrimp from July to September. White shrimp start appearing in August or September, and can be caught until December. If the water temperature drops below a certain point, the shrimp head south, or offshore, where warmer waters are to be found. When the shrimp migrate, the North Carolina shrimp boats either quit for the winter, or set traps for black sea bass.

Captain Jim and His Crew

Captain Jim was one of the best skippers in the Gulf shrimp fleet. His crew said that he had a sixth sense about shrimp. There would be nothing but miles and miles of water, and yet he somehow knew just where to drop the nets. Time and again, he would drop them exactly in the right spot to land a big load of shrimp. He and his crew consistently filled seven eighty-pound baskets a night with beheaded shrimp.

Captain Jim came from an old Georgia family, and had been fishing most of his life, as had many of his relatives. When he was a kid, he used to catch pan fish in a beach seine by walking one end of the net offshore in shallow water, encircling the school, and then walking the net back to shore, dragging the school with it. During prohibition, he gave up fishing for moonshining and running a gambling casino.

Captain Jim liked things really neat on his boat. He insisted that even the big diesel be utterly spotless. When he walked past it he would take out a white handkerchief and run it along the engine. If that handkerchief showed even the least bit of grease stain, he would chew out his crew and make them clean it again.

Jim had some unusual customs on his boat. He ordered each of his crew to come aboard each trip prepared with a riddle. The other crewmen kept trying to guess the answer. At the end of the trip, everyone was supposed to tell the others the answer to his riddle, but some guys never did.

Jim spent his winters shrimping off the western coast of Florida. When spring came he'd sail to Brownsville or Kingsville, Texas, or down to Campeche, Mexico and shrimp there.

Sometimes Jim's crew would take on private contracts from a buyer for a few hundred pounds of some type of fish. For instance, Spencer, the header on Captain Jim's boat, used to bring back 300 pounds of bonito every trip. They were used for catfood, and were worth six cents a pound. They were very simple to catch. Spencer would wait until all the trash fish, netted in the trawl along with the shrimp, were swept overboard. This would attract bonito by the hundreds, as well as many other fish. Then he would take a gaff, bait its hook with a few shrimp, and dip it in the water. In a few seconds a bonito would bite the hook, and he'd flip it on board. It took only a few minutes for him to get his 300 pounds. He'd chop their heads off, throw them in the ice hold, and deliver them to the buyer when he reached port.

Bonito are tasty food for humans as well as cats, but the fish are hard to dress, and perhaps this is why they have little market value. Greenback bonito have delicious white meat on their bellies and backs, but on their sides are strips of black meat, about two inches wide, that must be

Hauling a crab pot.

cut out. This operation takes time, perhaps too much time for them to be of commercial value.

Captain Jim didn't allow a drop of whisky on his vessel, so his crewmen were forced to do their drinking during their four-hour shore leave. Arnold, the rig man, usually tried to cram eight hours of drinking into one four-hour stretch. Jim and Spencer, the other crewman, would find him passed out in the park when it was time to leave. They'd just grab his arms and shanghai him aboard. He wouldn't be much good for a day or so. For the first day out, Spencer would have to be rig man.

Arnold was a damn good rigger, though, as long as he was sober. He could take a net that had been holed by a 500-pound shark and sew it up fast, so that every square of mesh was exactly the right size and shape. When he got done, you couldn't tell that the net had ever been damaged.

Once, the boat scraped against a dock, and a piece of the oak cap rail broke off. The broken piece had a compound curve in it, but that didn't bother Arnold. He just took out his machete and whittled away on a new piece of wood. It wasn't long before he had carved out a replacement for the broken piece, compound curve and all. After he fastened it in place and sanded it and polished it, the rail looked like new.

Arnold was a conch (pronounced "conk")—a man raised in the Florida Keys. He too had fished all his life. When he was a kid, he used to set a fine-meshed net in between two reefs. When the tide went out, shrimp would entangle themselves in it. The incoming tide would bring more shrimp into the net.

While Arnold spent his leaves catching up on his drinking, Spencer spent his time on shore catching up on something else. He would take a bucket of shrimp and a few lobsters, and give them to certain ladies in exchange for what they had to offer. One lady—he called her his "moon lady"—was just wild about his lobster.

Crabs

Winter comes to Chesapeake Bay sometime between October and December. It is often preceded by an unusually warm day, after which the dry gales make the bay rough and hazardous. Sails of yachts get torn right off the mast during the gales. Experienced workboat captains stay in port when the gales hit.

The gales eventually stop, leaving the air cold and the water crystal clear. The water temperature drops into the 50s, and the jimmies (male crabs) leave their shallow-water haunts in search of deeper and warmer water. Since they are cold-blooded, low temperatures slow down their metabolism and make them sluggish. They search for warmer water so they can be active enough to forage for food. Most fishermen put away their crab pots at this time of year, and get out their dredging gear.

Snow starts to fall sometime in December. The fishermen claim that the crabs go dormant the day snow hits the waters of the bay, as if someone turned off a switch. Scientists say it is when the water temperature drops below 40 degrees that the crabs become dormant.

When the cold weather hits, the crabs bury themselves

in the mud of the bottom, digging in backwards so that only their antennae and eyestalks are showing. Unless disturbed, they remain in their cozy burrows, barely moving, throughout the winter.

Dredgers

Due to Virginia's special season that runs from December to March, many of the dormant crab are disturbed. Power boats tow heavy bags made of chain and twine along the bottom during these months. These dredges dig into the mud as they are dragged along, scooping out the buried crabs. Many people, fishermen included, claim that this dredging is tearing up the bottom too much, and will soon ruin the fishery. The dredges weigh twice as much as the oyster dredges that are also used on the bay. Only Virginia allows these heavy dredges. Maryland has always been more strict in its fishery regulations, and this type of dredging is prohibited in its waters.

The seventy or so vessels that dredge for winter crab are the largest and most powerful commercial fishing vessels on Chesapeake Bay. The first of these vessels were made from old schooners by chopping off their masts and fitting them with large engines. Many of today's dredgers still retain the graceful schooner lines.

Many of the dredge boats anchor in Cape Charles, near the southern tip of Virginia's eastern shore. None of the crabbers live in Cape Charles, though. It is a ghost town that was once the terminal port for the Cape Charles–Little Creek ferries, a vital link in the New York to Florida coastal water route. Many of the crabber's families live on Tangier Island. The crabbers "drudge" six days a week, returning to Tangier every Sunday for the gospel service.

The boats dredge at two or three knots. The tined bar at the mouth of the dredge digs into the sand and mud of the bay's bottom, forcing the dormant crustaceans from their lairs. The dredges stay down for ten to twenty minutes, during which time the two crewmen and captain catch a bit of rest. At the end of a "lick," the dredge is hauled. The tow chain sets up a deafening clanking as it is pulled aboard, waking up any crewman who has managed to fall asleep.

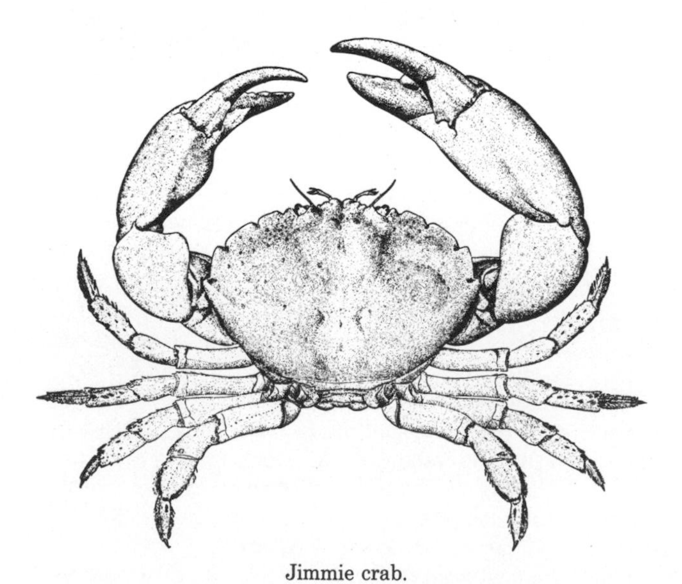

Jimmie crab.

When the dredge surfaces alongside the boat, one of the crew takes the turning stick, a skinny log made of loblolly pine, jams it between the struts of the dredge, and rotates the dredge so that the open part of it is against the boat. The dredge may be shaken up and down a bit to free it of mud and weeds. Then it is hoisted onto the deck, guided by the crew. The men open it up, and the crabs fall on the deck.

Every four dozen crabs make a basket. Three baskets make a barrel. Twenty-five barrels (the limit for a day) sell for two to three hundred dollars during the winter.

After the dredge is emptied on the deck, it is immediately lifted by two men and dumped over the side, so that no dredging time will be wasted. The boat turns around at this point to do another lick, and the waves hit it broadside, making it "roll like twenty drunken whores." It is easy to fall overboard, and dangerous handling the heavy dredge. After the dredge is returned to the water, the crewmen cull, or separate, the jimmie crabs from the sooks (females). When the day is warm, the crabs are restless and active, and try to bite the crew as they are culling them.

Dredging ends sometime in March, when the female crabs "brighten up their fingernails." When their claws start turning redder than usual, the fishermen know that they will soon leave their winter burrows, after which they can once again be taken in pots.

Dungeness Crab

Dungeness crab, a variety taken over much of the West Coast, seem to appear in huge numbers every six or seven years, then slack off during the in-between years. The winter of 1975–76 was one of the best years on record. Some California boats grossed $300,000 during that season. The last good season before that was in the winter of 1968–69.

It is a mystery why excellent seasons occur at regular intervals. Baby crabs that have recently been spawned float on the surface for a while, drifting with the currents before they settle to the bottom. If the currents carry them into water deeper than 100 fathoms, the high pressure crushes them when they sink. If they are carried into shallow water, they are able to mature. Perhaps certain currents recur every few years, and it is these currents that are responsible for good and bad seasons.

Dungeness crab pots are round in shape, and have trap doors that let the crabs in but not out. Their frames are made of round steel stock welded together. The webbing is woven from stainless steel wire. Rubber insulators are placed between the webbing and the frame to prevent electrolysis corrosion. The frame is wrapped with rubber strips cut from inner tubes. These rubber strips are easier on the hands than hard steel. Also, the pots bump against and scrape the side of the boat as they are hauled out of the water and the rubber prevents damage to the boat.

An efficient crew can haul, rebait, and reset 300 pots in one ten-hour day. The pots sit on the bottom and are con-

Hauling a Dungeness crab trap.

Fisherman with Dungeness crabs. *(Rochelle Smith)*

nected with polypropylene line to buoys floating on the surface. There is a buoy for each pot. Pots are laid out in rows, or strings, with as many as 100 pots to the string. The crabbing boat will steam along parallel to a string. When the boat comes up alongside a buoy, a crewman will grab the line connected to it with a boat hook, haul buoy and line on board, and slip the line into the hydraulic pot hauler. The pot hauler is mounted on a boom. As the pot is pulled from the water the boom is raised, making it easier to swing the pot over the rail.

The pot is emptied of its catch and rebaited. The boat never stops moving during the hauling operation. About seventy-five feet before the next buoy is reached, the baited pot and buoy are pushed overboard. This is a dangerous time, for if a man gets his foot caught in the swiftly unreeling line connected to the pot, he can be dragged overboard.

Vessels usually work into the wind, so that after a string of pots has been hauled and reset, the boat returns downwind before starting on the next string, giving the crew a few minutes of rest.

Alaskan King Crab

One of the newest and most hazardous of North American fisheries is the Alaskan king crab fishery. It is an extremely rich fishery and many millions of dollars have been made from it.

Since crab must be delivered live to the buyer's, boats must be equipped with tanks of seawater to keep the crabs in. The water sloshing back and forth in the tanks tends to make a boat unstable, especially during the hundred-knot winds that are frequent occurrences on Alaskan waters in wintertime. The heavy pots, weighing up to 800 pounds apiece, add to the stability problem.

Fishing Vessels

Trawlers

Trawling is the most common method of commercial fishing in the world, originating in Europe in the late eighteenth century. Most of its development came in England in the nineteenth century. In simplified terms, a trawl is a large funnel-shaped net dragged behind a fishing vessel. There are several types of vessels designed especially for trawling, ranging in size from 50-foot shrimp trawlers to 295-foot factory trawlers like the *Seafreeze Atlantic*. Some trawlers are designed to haul their nets in over the side, and others bring them in over the stern. The sterns of some of these last types of trawlers end in ramps, up which the trawls are dragged.

The first American trawler was built in Essex, Massachusetts in 1891. She was an 85-foot yawl, a two-masted sailboat with the main mast toward the bow and a short mast, called the mizzenmast, near the stern. The design was copied after English boats, and her crew was also English.

It wasn't until 1904 that the American trawling industry really got going. That was the year that a group of Boston businessmen imported a set of plans from England and built the *Spray*. The *Spray* was such a successful trawler that a whole fleet of similar vessels was soon built.

The *Spray* was 136 feet in length, had a 4500 cubic foot fish hold, and was steam powered. By 1926, 43 vessels of the *Spray* class were operating. They were slightly larger than the *Spray,* and fitted with Scotch boilers to run the steam engines.

Gas and diesel engines were coming into use at this time. By 1920 a fleet of converted schooners, sailing vessels with two or more masts rigged fore and aft, were trawling. A few years later, several vessels designed especially for use with the new engines were built. One of these vessels, the *Mariner*, had a combination diesel and

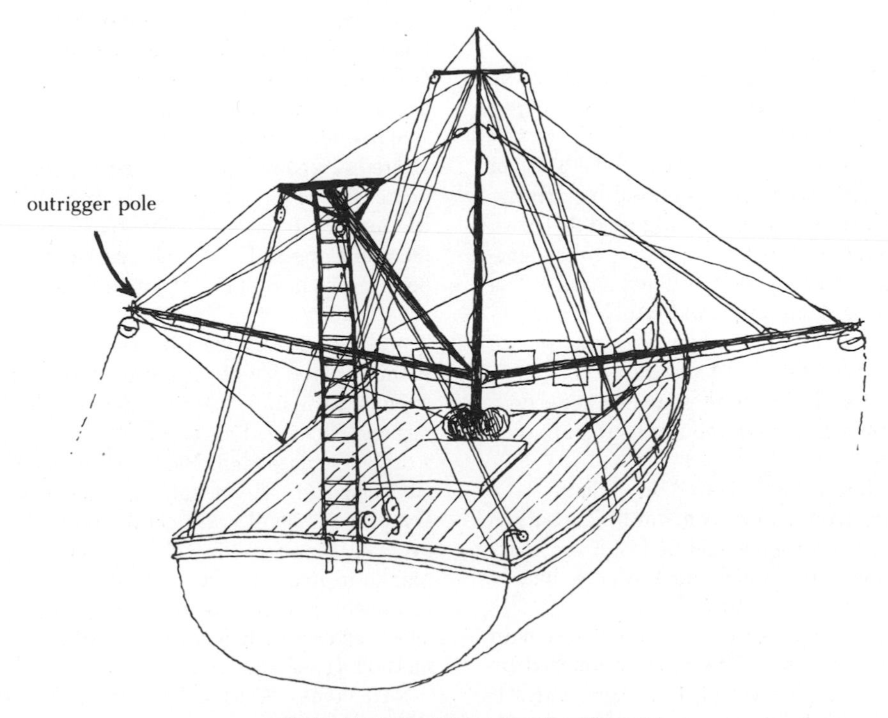

Shrimp trawler. *(John Douglas Moran)*

electric motor. This was an experimental engine, and was never widely used.

By 1930 diesels far outnumbered steam trawlers. After World War II the United States Army bought many of the large trawlers for use in Germany. This depletion of trawlers in America added to the postwar construction boom, for the trawlers that were sold had to be replaced. In 1947, though, high construction costs and falling fish prices brought construction of new trawlers to a virtual standstill.

Although schooners have disappeared from the trawl fisheries, their lines have been retained in many of the smaller trawlers. These schooner-type trawlers still have two masts and use a small riding sail.

Many of the vessels in today's American and Canadian trawler fleets are old and outdated, but still must be used, for the money is just not available to replace them. Of the new vessels that have been built, the largest are the 295-foot stern trawlers *Seafreeze Atlantic* and *Seafreeze Pacific*, built in 1968 for American Stern Trawlers, Incorporated. They have fish holds almost ten times the size of the one on the *Spray*. Most of the other new vessels are quite a bit smaller, between 100 and 160 feet in length.

Canada has one of the most varied trawler fleets in the world. Her vessels differ widely in style, layout, cost, hold capacity, cruising range and overall size. Most of her vessels are not refrigerated, but ice their catch. This is because Canada is near some of the finest trawling grounds found anywhere, like the Grand Banks of Newfoundland, the Nova Scotian Shelf and the Gulf of St. Lawrence. A boat can leave port, load up on cod and haddock, and be back within a week. Thus, Canada doesn't have much need for refrigerated trawlers that keep fish fresh for months.

Shrimp Trawlers

The design of the Florida shrimp trawler is based on the lines of Greek sponge boats. Greek fishermen brought their boat designs and construction methods from the Old Country when they came to Florida. Many of the boats in the Florida area are still owned and operated by Greeks.

This type of shrimp trawler is now being used in Texas, Mexico and South America. It is typically fifty to seventy-five feet in length, and made of oak, cypress and fir. The deckhouse is forward and the mast and winch are amidships, leaving a large work space in the after part of the boat. Years ago, the trawl nets had to be hauled by hand. Thus, large crews and a sizable work space were needed. The old design has remained, even though trawls are now hauled by winches.

Shrimp boats are fairly standardized in design, for a company will often transfer a crew from one of its boats to another, and valuable work time would be lost if the boats were all different from each other and the crew had to spend time adjusting to the new boat.

The boats are designed so that the deck area will remain dry, even in a fairly rough sea. This is accomplished by building ample sheer into the vessel, both fore and aft. This means that the boat's hull curves noticeably upward, both near the bow and near the stern. Shrimp trawlers are built with more sheer than most other boats, and have been criticized by some designers, who feel that the amount of sheer is excessive, and is more for show than for function.

Shrimp trawlers used to drag for shrimp with one big trawl net. Now, most of them have switched to two smaller trawls, one trailing from each side of the vessel. The two small nets are supposed to fish more efficiently than one big one because they are better able to adjust to irregularities on the bottom. Each trawl is attached by a long line to a boom, called an outrigger, that hangs over the side of the boat. The two booms are often made out of old oil field drill stem pipes. The Texas oil companies use twenty-four-foot sections of heavy walled four-inch pipe in their operations, and discard them when the wall thickness is worn down to half an inch. These discarded pipes make excellent and inexpensive outriggers for the shrimp boats.

A sixty-seven foot trawler can carry twenty tons of fresh, well-iced shrimp. Some trawlers will have the galley and captain's quarters in the deckhouse, and the crew's quarters below decks in the forecastle (the part of the boat right behind the bow). A new trend in shrimp boats is to build the sleeping accommodations for the whole crew above decks, in the house.

The galley area in these boats is quite roomy compared to other fishing boats of this size, and on many vessels is attractively finished with cypress paneling. Extra comforts like these can help morale on a long trip. The boats are quite a contrast from the old Monterey lampara netters, where "you didn't sit down during the daylight hours, and if you wanted to take a crap, you used a five gallon bucket and dumped it over the side."

One man often owns a fleet of half a dozen or so shrimp boats. This is quite a different situation from many other fisheries, like the California salmon fishery, where the captain of the boat is usually the owner as well.

Many shrimp trawlers are assembly line produced. There are about a dozen stages of construction, each one being carried out by a separate crew. The first operation is the laying of the keel. The keel is a nine-by-twelve-inch piece of Douglas fir or pine that runs the full length of the boat. In the next stage, the molds are set up that will guide the framing crew. The framing crew steam bends the boat's white oak ribs into the proper shapes, and nails them in place. The framers are followed by the structural crew, who install the deck beams, floor timbers, engine bed and decking. This is the largest crew and is made up of a dozen or more men. Next, the planks are fastened to the ribs, enclosing the vessel from the deck down to the keel. Douglas fir planks are used above the water line, and cypress is used below. No calk is squeezed between the planks to make the boat watertight. When the vessel is launched, it will leak for a few hours until the planks swell, after which it will be watertight. This no-calking method saves time and materials.

Next come the sanding crew and first painting crew, who apply a coat of primer to the boat. After this, a crew

120-foot Washington trawler-crabber. *(Bob Carver)*

of skilled finish carpenters build the deckhouse, panel it and make it into an attractive and comfortable place in which to live. Next come the mechanics, then more painters, then the machining crew, and finally the welders, who install the mast, the outrigger booms and the mufflers.

A 67-foot shrimp trawler will carry 6000 gallons of fuel, 60 gallons of lube oil, 650 gallons of fresh water and 45 tons of ice. It is made to cruise at three knots, but can cruise at eight if it has to.

Jonesporters and Novies

Maine lobster boats are often called Jonesporters, after the town of Jonesport, Maine, where early boats were made. Most Maine lobster boats come from little backyard shops that build perhaps two boats a year, usually during the winter off-season, for the boat builders are often busy lobstering during the summer months. No blueprints are used in the construction of these boats. Builders use carved wooden models that are often decades old. They take the measurements off the models and scale them up to the size of the boat they are building. This method gives each yard's boats a distinctness and individuality.

One Beal's Island builder has built over one hundred lobster boats in his backyard shop. They average thirty feet in length, and it takes him three months to complete one. The man thinks his boats are better than the ones produced at Jonesport. But then, as he says, "Every crow thinks his feathers are whiter."

Most Jonesporters used to have only a canopy for shelter. Now, they are all built with cabins. A few lobstermen have fiberglass craft, but not many. Maine fishermen follow their traditions more rigidly than fishermen in any other region of the United States. Most Maine lobstermen prefer the tried and true oak-framed, cedar-planked lobster boats.

Novies are craft made in Nova Scotia. They are known for their high bow and low stern, and are the most popular inshore lobster boats in Massachusetts waters, as well as in Nova Scotia. The high bow is for breaking through waves on a rough sea. The low stern was developed during the days of hauling pots by hand. It was easier to lift the pots from the water to a deck with a low stern.

The main advantage Novies have over American boats is that they are cheaper. They can be bought for roughly half of what an equal-sized American boat sells for. In 1973 a new Novie cost somewhere around $8000, whereas a new Jonesporter was about $14,000. One reason for the Novie's low cost is that the planking is not screwed on as it is on American boats, but is clinch-nailed. A carpenter on the outside of the hull drives a nail through planking and frames, and a carpenter inside merely bends the ends of the nails over. The nails start coming loose after a few years, and must be replaced with screws. This possibly offsets the initial low cost of the boat.

Maine Jonesporters are similar in appearance to Novies, for they too have high bows and low sterns, but their hulls are narrower and shallower, making them considerably faster and more useful if the lobsterman must travel long distances to and from his pots.

Inshore boats (boats meant for work within several miles of shore) average twenty-eight to thirty-five feet in length, and have a beam (width) of eight to ten feet. They are generally built of cedar planking over oak frames, and are powered by converted automobile engines or special marine gasoline engines, although diesels are becoming more common. Diesel engines have lower fuel and maintenance costs, but higher initial costs.

If the boat is fishing in winter, a temporary plywood cabin might be built behind the small permanent cabin, in order to give the fishermen more protection from the weather. The plywood cabin is removed during the warmer months.

Pots being hauled from the water bang and scrape on the hull, and would eventually damage it if chafing gear weren't attached to the hull. Chafing gear consists of oak strips fastened to the hull in spots that receive a lot of abrasion. Fiberglass skid plates are also used for this purpose.

Many boats have a mast near the stern. During rough weather, the steadying sail is hoisted onto it. This sail helps keep the boat headed into the wind during hauling operations.

Offshore Lobster Craft

Boats in the new offshore fishery are bigger and more rugged than inshore boats, for they have to stay out longer and weather rougher seas. Since it sometimes takes twelve or more hours to reach the fishing grounds, the boat must have a big enough fish hold to make the long trip profitable.

Independent lobstermen do fish the offshore grounds, although they can't catch as much as the hundred-foot-long, steel-hulled corporation boats. A lobsterman's offshore boat might be fifty-five or sixty feet in length, and would easily have cost $100,000 if he bought it new. It is similar in layout to the inshore craft but built for heavier duty. The hydraulic pothaulers can lift two tons, instead of only 800 pounds, like the inshore crafts' haulers. The finest depth sounders and radios are used, and traps are bigger and stronger. Inshore vessels are usually close enough to shore to navigate by sight, but offshore vessels need expensive radar and loran sets to find their way.

The Alaska Limit Boat

Many large and capable boats from Washington, Oregon and California invaded Alaskan waters in the late 1930s to fish in the state's abundant salmon grounds. Alaskans couldn't afford to build boats the size of these invaders. Their small boats weren't able to compete with the large out-of-staters, and the Alaskans were in danger of being run out of their own fishery. They banded together and got a law passed that banned boats of the dimensions of those owned by many Washington fishermen from purse-seining salmon in Alaskan waters. But the Washington fishermen discovered that with slight modifications many of their boats would meet Alaska's requirements. The Alaskans retaliated by passing a new tougher regulation limiting the overall length of the boats to fifty-eight feet. By the time this law was passed, many Alaskans could afford bigger boats, but still didn't want bigger boats to be legalized, for then their old vessels, into which they had sunk large amounts of money, would be greatly devalued.

The rich Alaskan fisheries support many fishermen from the "lower forty-eight," as well as Alaskans. Washington fishermen are especially dependent for their livelihood on their yearly trips to Alaska. Thus, few Washington fishermen buy salmon boats larger than fifty-eight feet.

The irony of the Alaska Limit Law is that it has caused as many disadvantages to Alaskans as advantages. Limit boats, although fine during the summer salmon season, are of little value in the new and very profitable wintertime crab fisheries. The crab fisheries were developed

Alaska limit boat with seine skiff on stern. *(Marco)*

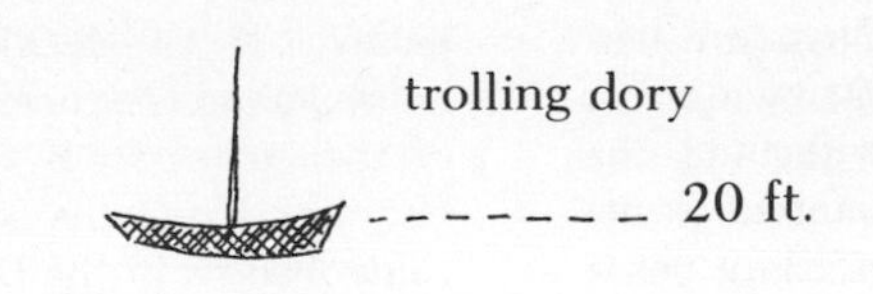

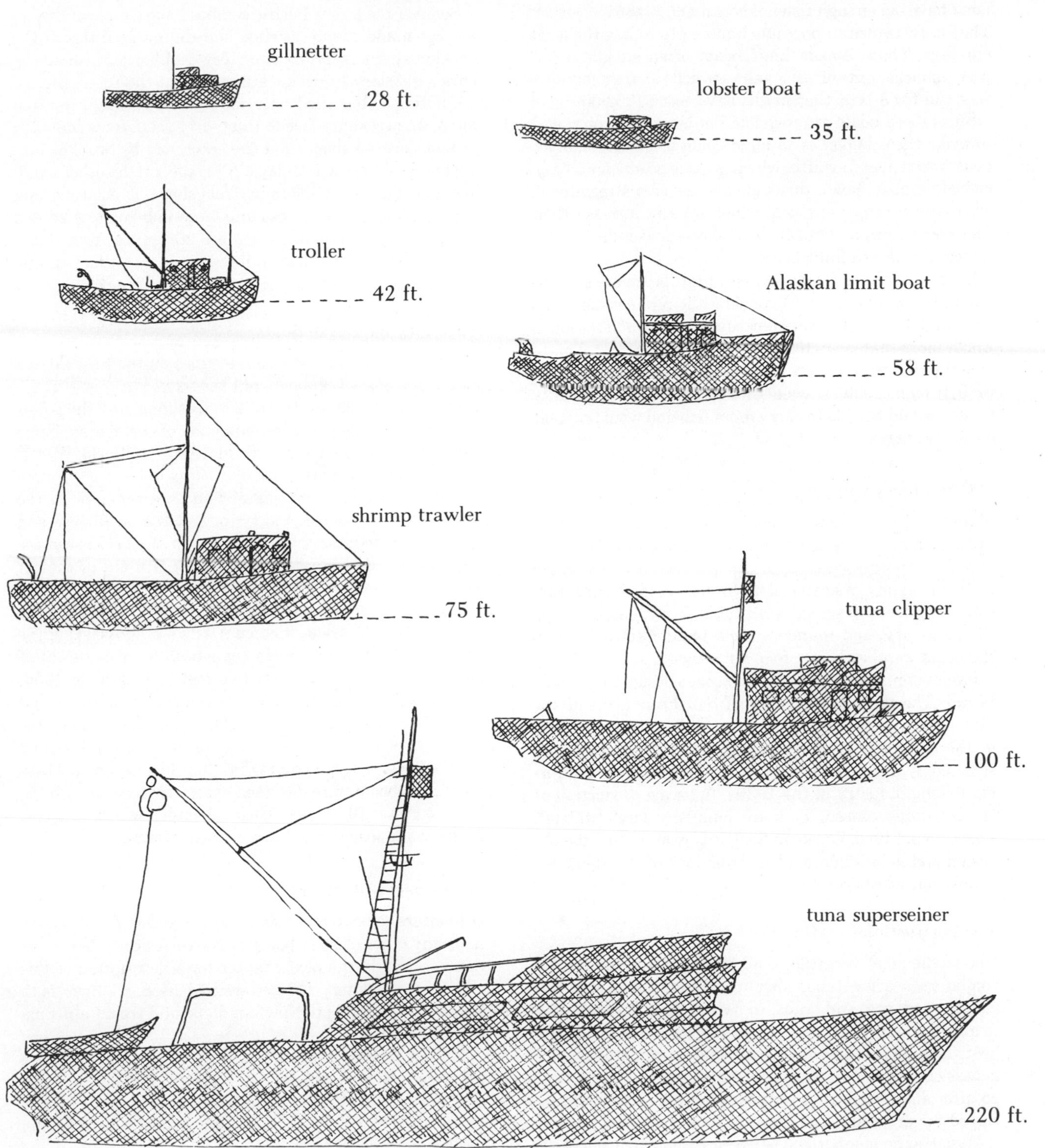

(John Douglas Moran)

after the 1938 law was passed. The limit boats are just too small to fish the open sea during an Alaskan winter. A ninety-foot crabber is more the size to withstand the hundred-mile-an-hour winds that make mountain peaks and Grand Canyons out of the ocean's waters. Limit boats don't have big enough fish holds to make long trips, either. They must return to port much more often than the large crabbers. Thus, Alaska limit boats often sit idle a full nine months out of the year, which is tremendously wasteful for a boat that might have cost $100,000.

Since these boats are restricted in length, the only way to make them bigger is to build them beamier. Some exceed twenty feet in width, which is quite beamy for a fifty-eight-foot boat. Boats this wide are not very streamlined, and require larger engines, which greatly increase their fuel consumption. Four-hundred-horsepower diesels are typical on Alaska limit boats.

In spite of these disadvantages, limit boats are still being built, for a buyer must consider the resale value of his boat, and a boat that can be used during salmon season is worth more than a slightly larger one that cannot. This is why the smaller crabbers and groundfish trawlers tend to be fifty-eight footers, even though sixty-five or seventy footers would be able to carry more fish and would be a bit more seaworthy.

The Monterey Clipper

Monterey clippers trace their ancestry back to the small, upswept-bow sailing craft that were used on the Nile River over a thousand years ago. The boats were later adapted by Sicilian and Corsican sailors for use on the Mediterranean. They were designed with high, flared bows to keep the decks dry, and rounded, tucked-under sterns to help the boats weather heavy following seas.

Italian immigrants brought the boat's design to the New World. The Beviacqua family of San Francisco and the Siino family of Monterey began manufacturing the boats in the early 1900s. Four hundred Montereys were constructed by the boatyards. With the decline in the California fishing industry in the 1930s, full-scale production of the Monterey ceased. Few are built anymore, although Puget Sound Boat Works in Seattle is resurrecting the old design and is building modified versions of the Monterey clipper out of fiberglass.

The Pacific Combination Boat

One of the most versatile of all fishing boats, the Pacific combo goes after tuna, shrimp, salmon, mackerel, anchovies, cod, sole and crabs, to name a few. It fishes year in and year out, changing its gear several times during the year. It might setline for black cod or trap fish for Dungeness crab in wintertime, troll for salmon in spring, and go after albacore during summer and fall. Or it might be rigged for dragging, and trawl all year for groundfish.

Most Pacific combos are between twenty-five and sixty feet in length, made of wood or steel, and powered by a gasoline or diesel engine. The deckhouse is in the forward part of the boat, with the engine room directly below. The galley is in the deckhouse. The crew sleeps in the deckhouse and in the forecastle. There is a flying bridge on top of the house, from which the skipper can spot marker buoys or direct the set of a purse seine. The boat can be controlled from the bridge or from the wheelhouse.

Some of the larger Pacific combos have a separate sleeping cabin and roomy kitchen and dining facilities in the deckhouse. Some have a two-story deckhouse, with all the crew's quarters on deck, rather than in the forecastle.

Pacific combos used only for salmon fishing are often fairly small (twenty-five to thirty-five feet), for salmon are caught close to shore, and the catch can be brought into port and sold every few days. Albacore, on the other hand, are caught as far as 300 miles from shore. This means long trips, and long trips necessitate large fish holds. Also, the sea gets very rough when you are 300 miles from shore. Albacore boats are thus quite a bit larger than salmon boats. Most are at least forty-five feet in length.

Steamers and Purse Boats

Diesel engines replaced steam engines on the East Coast's menhaden boats years ago, but the name steamer still lingers on. The boats are 100 to 200 feet long, and the larger ones can carry over a million pounds of menhaden. Some steamers have been made from 138-foot World War II wooden minesweepers.

The diesel engine or engines on steamers are in the stern, the fish hold is amidship, and the deckhouse and crew accommodations are forward. In the deckhouse are large pumps that are used to suck the fish out of the net and into the hold.

The two purse boats—the smaller craft that handle the purse-seine net—are carried on davits (see glossary) in the after part of the ship. Purse boats used to be made out of steel, and were about thirty-two feet in length. In 1956, thirty-six foot aluminum boats began replacing them, and today few steel boats are left. The aluminum boats have proved to be lighter, easier to handle and more stable, and yet the boats weigh no more than the old steel craft. These aluminum boats were designed especially for use with the Puretic Power Block (see "Gear" chapter), which require a larger work space than the old boats offered.

West Coast Gillnetters

Gillnetters come in more shapes, sizes and styles than perhaps any other class of boats in North America (see "Gillnets" section in the next chapter for a description of this method of fishing). Gillnetters are used all over both coasts, from Florida to Newfoundland and from California to Alaska.

The archetype gillnetter on the West Coast was the Columbia River sailboat. A beefier version of this craft—the Bristol Bay gillnetter—was used on Alaska's Bristol Bay until 1951, when power boats were finally legalized on that bay.

The Columbia River boat eventually became the Columbia River "bowpicker," a boat still used on the Columbia, and in Washington's Willapa Bay and Gray's Harbor. The

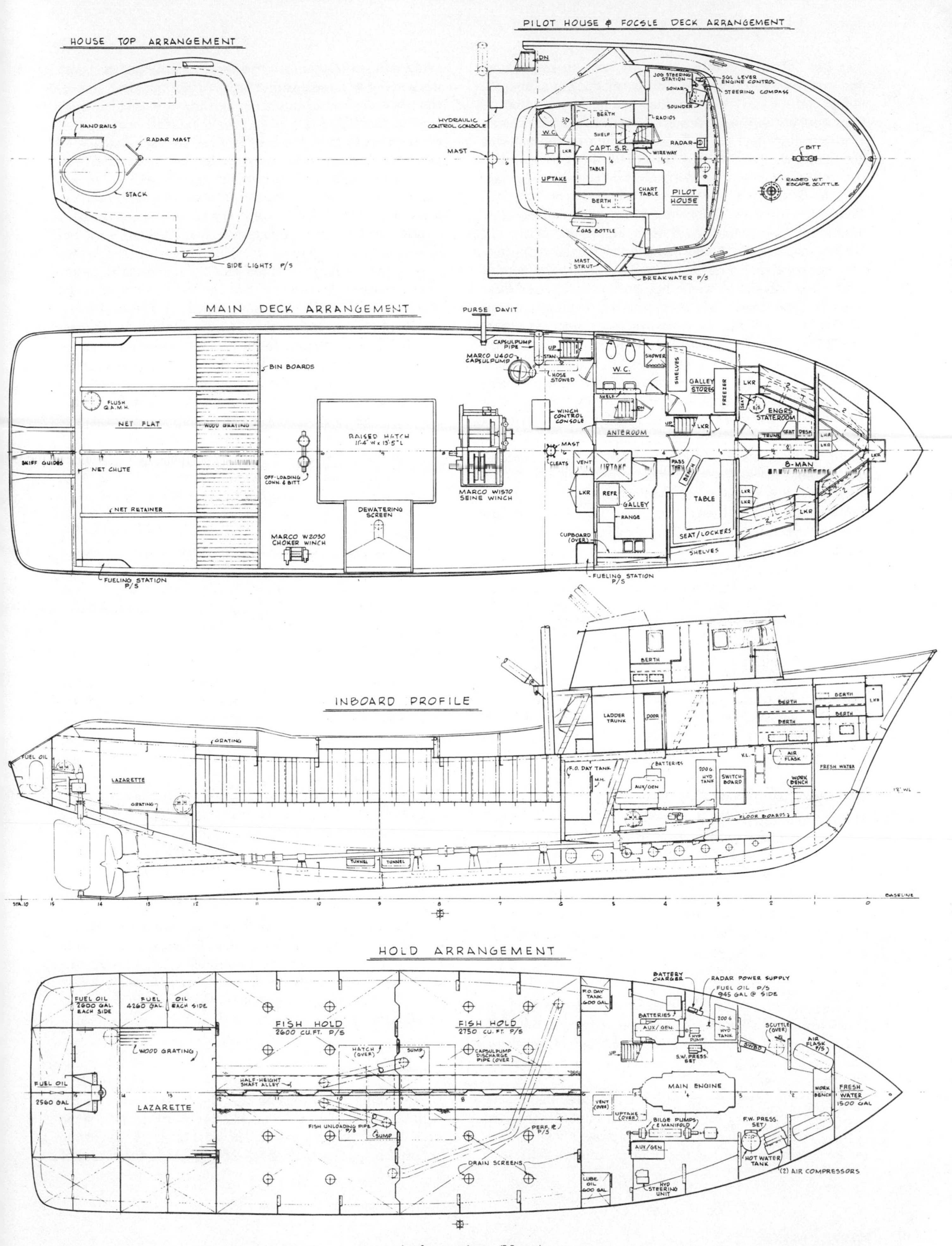

Anchovy seiner. *(Marco)*

boat gets its name because the work space is in the bow, and it is from there that the fish are picked out of the gillnet. When engines were first installed in the old sailboats, they were put in the stern, so they wouldn't get in the way of the sailing gear. Setting and hauling of the net was done from the bow, so that the net wouldn't be fouled in the prop. This bowpicker design has persisted until today. The modern hydraulic-powered net haulers are still located in the bow. The old double-ender sailing craft design has been modified, though. Both ends of the old boats came to points; the present-day power boats now have square sterns. The boat is well suited for the areas in which it is fished, but would not work in most other locations. It is too small, and the crew accommodations are too spartan and uncomfortable. It is a short-range day boat, and isn't very useful in areas where the fish are far enough away that overnight or several day trips must be made.

When power boats were legalized on Alaska's Bristol Bay, many new designs were tried. Today's Bristol Bay gillnetter is perhaps not as graceful as the old twenty-eight-foot double enders, but it is far more efficient. One popular type measures thirty-two feet in length, is diesel driven, and has a fiberglass hull and wheelhouse. There were about 65 of these craft operating on the bay in 1975, and around 145 of them throughout Alaska. Some are used for swordfishing, or in several different fisheries during the year.

One of the smallest fleets of gillnetters is found in Chenois Creek, in Washington. None of the nine boats in this fleet is more than twenty-four feet long. They are powered by outboard motors, and have little "doghouses" aboard to shelter the crew. The doghouses are sorely needed, for the boats fish at night and many of the nights are very damp and very cold. All the boats are designed with quite shallow drafts so that they do not run aground in shallow Chenois Creek. They are a good example of boats that are excellent in the small fishery for which they are designed, but are of very limited use elsewhere.

East Coast Gillnetters

The Florida mullet gillnet skiff is a good craft for gillnetting in placid, sheltered areas such as in the inland waterways of Florida, Georgia or Texas. The boats are twenty feet long and are built with plywood planking.

The small motor, usually twelve to twenty-five horsepower, is mounted in a well in the forward part of the boat. The forward location of the motor has two advantages: it leaves a large work area in the stern, and it prevents the nets, which are hauled over the stern, from fouling in the propeller.

Gillnets are used in many of the deep rivers on the East Coast, such as the Delaware, the Connecticut and the Hudson, to harvest fish on the way to their spawning grounds. The nets are strung between two poles which are driven into the riverbed.

The tides in these rivers often exceed three knots, and the water levels vary enough so that at low tide, the tops of the gillnets are often right at the surface. A river gillnet tender, quite similar to the Florida mullet skiff, has evolved over the years to tend these nets. When the boats were powered by two pairs of oars rather than by a motor, the usual method of approaching the net was stern first, against the tide. After reaching the net the two oarsmen rowed hard to hold the boat against the net. The one or two fishermen on board stood in the stern, hauled up sections of the net and picked the fish out of them. After one section was picked clean, the fishermen would haul the small boat along the net, while the oarsmen kept the boat heading straight and prevented it from drifting with the tide.

The oarsmen have been replaced by an outboard motor, mounted near the bow to give the men plenty of work space in the stern. To keep the propeller far from the net, the boat still approaches the net stern first. The mounting well for the motor is made out of a section of heavy-gauge pipe. The driveshaft runs down through this pipe into the water, and the motor itself is clamped onto the pipe.

One of the most extensive uses of outboard craft is in the salmon fisheries of Alaska and the Pacific Northwest. The typical salmon net tender is a very sturdily constructed boat of about twenty-six feet in length. The outboard motor is placed in a well near the stern. The well is often on one side of the boat in order to allow a bigger work space in the stern. This boat has a much heavier frame than the river gillnet tender. The framing members on the river gillnet tender are two and a half feet apart, whereas on the salmon net tender, there is a framing member every fifteen inches. Its strength is needed in order to withstand the rough waters of such places as Cordova Bay, Alaska.

Superseiners

Everything about a tuna superseiner is large. Its size—up to 300 feet. Its purse seine—some are a mile long. Its carrying capacity—the *Margaret I* holds four million pounds of tuna. Its engine—3600-horsepower, twenty-cylinder marine diesels are common; and price—the largest cost five million dollars.

Some have landing platforms for helicopters on them. The electronic equipment on one boat includes a gyrocompass, a radar set, two ADF radios, a depth sounder/recorder, three single sideband radios, two VHF-FM radios, two CBs and a weather facsimile machine. It also has a Triton water maker on board that produces forty gallons of fresh water an hour.

The captain's quarters on these boats are nothing short of luxurious. Black-walnut-paneled living rooms, soft leather furniture, heavy walnut tables, large rooms, attractive paintings on the wall. Some of these quarters look more like fashionable penthouses than they do cabins on board a fishing boat.

The Seine Skiff

Purse-seining operations on the West Coast, whether for Southern California tuna or Alaskan salmon, use a seine skiff to help the main vessel. The skiff takes one end of the net, and helps the mother vessel to encircle the school of

Superseiner. *(Campbell Industries)*

fish with it. Also, as the net is being pursed and hauled, the skiff helps prevent it from drifting under the seiner and fouling its prop. During good weather, the skiffs are sometimes towed behind the mother vessel. During bad weather they are hoisted aboard.

Skiffs used to be built of wood, although they are fast being replaced by those made of lightweight metals like aluminum and Corten steel. Typical lengths are seventeen to eighteen feet. They are equipped with fairly large inboard diesel or gasoline engines for their size. They need this power, because they are used basically as small tugboats, and must be able to pull considerable loads.

Tuna Clippers

The thing one notices right away about tuna clippers is how low their sterns sit in the water. The main deck, in fact, is practically level with the water. This "low freeboard" makes the work of hauling tuna from the water much easier. The fishing is done with poles and lines from platforms mounted outside the main deck rails. When the tuna bite the hook, they are flipped out of the water, over the deck rail and onto the deck. This operation would be much harder if the deck and stern were higher than they are.

The low stern looks like it would allow a vessel to be easily swamped, thus making it unseaworthy, but this is not the case. Tuna clippers roam all over the Pacific, and are habitually subjected to very rough seas. One thing that does make a vessel unstable is having too much weight high up, as early tuna clippers had. Large tanks filled with seawater and live bait were mounted on the decks of these early clippers. At least two of these boats capsized, and a third was lost with all hands. Today, due to the above catastrophes and due to pressure from the companies that insured the boats, strict stability tests must be performed on all tuna boats.

The boats are 50 to 150 feet in length, and carry crews of ten to twenty men. The crew's quarters are generally in the deckhouse, and their rooms are as wide as the deckhouse is, so that portholes can be opened on both ends of the rooms, allowing good circulation of cooling air. On top of the mast is a covered crow's nest, where the captain or lookout sits with a pair of binoculars, watching for signs of tuna.

Many of the boats are painted in a uniform style. The hull and deckhouse are painted white, the boxes and working areas gray, and the wooden rails that run the length of the boat are stained with a natural finish, to emphasize the vessel's long lines.

Early clippers cooled their catch with crushed ice, and the crews of these boats had to work well into the night icing the fish in the hold. In the 1930s chilled brine systems began replacing icing. In this system fish were put into sealed wells filled with brine and kept there until the boat reached port.

There once was a skipper who caught an extra full load on one of his trips, and was worried that the seawater

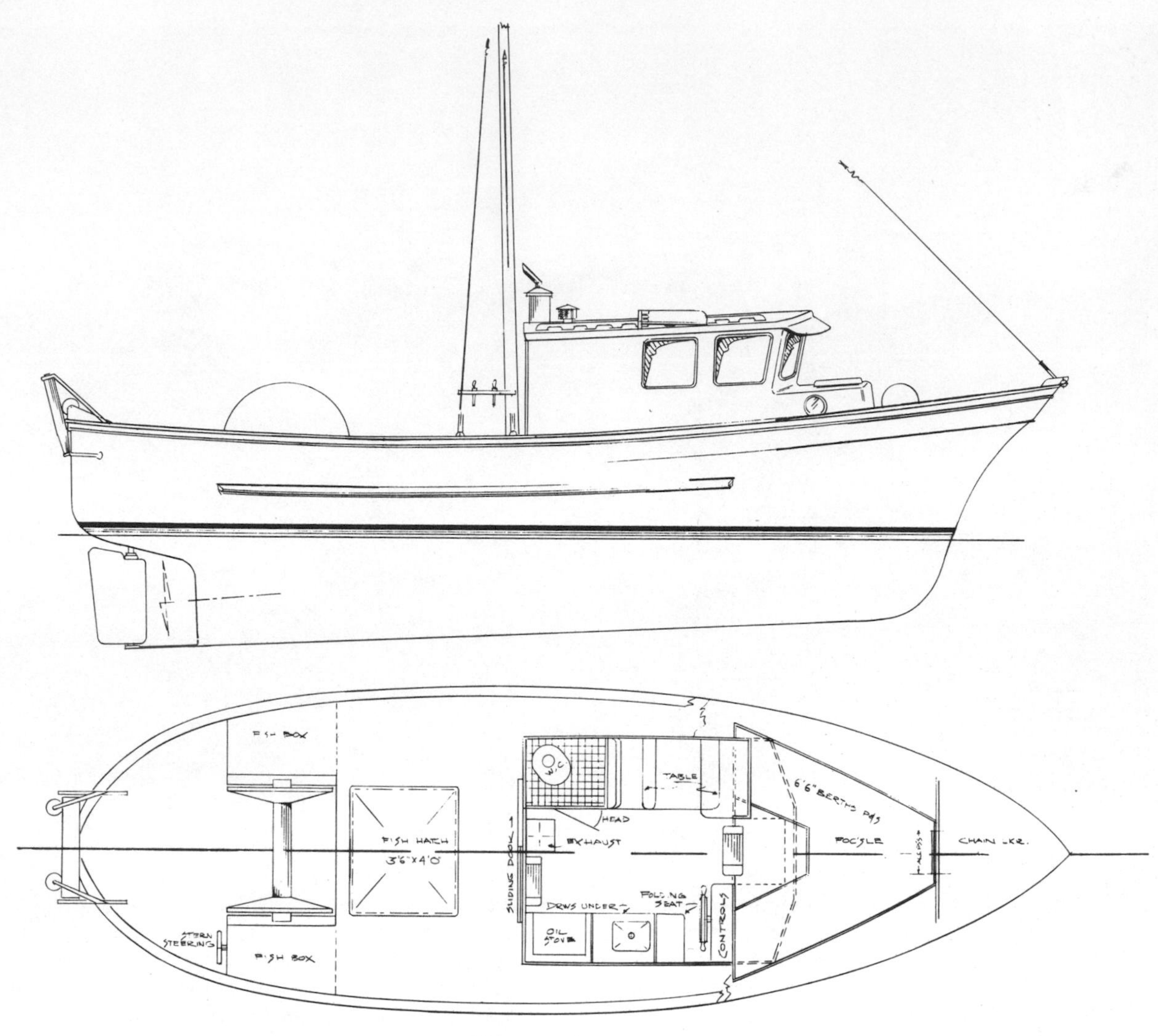

Gillnetter. *(Marco)*

sloshing around in all of his fish wells would capsize his boat if he hit rough weather. He decided to pump the brine out of a few of the tanks that had already been refrigerated, hoping that the chilled air in the well would keep the fish from spoiling. The vessel was considerably lighter with its drained tanks, and reached port in very good time in spite of some bad weather. When he opened the wells, he found that the "dry" fish were in excellent condition, perhaps better than those in the brine-filled wells, where the fish had absorbed a certain amount of the salt contained in the brine.

The captain repeated his experiment on several more trips, each time with good results. Soon after this, the whole fleet adopted his method of first chilling the fish in brine (which lowers their temperature much faster than air chilling), then pumping out the brine, and keeping them cool with chilled air. This is the method still used by the American Tuna Fleet.

Many of the fishermen are deeply religious, and put small chapels in their boats. The chapels are decorated by the fishermen's wives with fine lace and expensive material, so that the men may have a breath of their homes and families, even though they are thousands of miles away.

Outboard Craft

There are a number of small fishing boats that are powered with outboard motors, instead of built-in gasoline or diesel engines. Some were mentioned in the "Gillnetters" sections of this chapter. These craft are under thirty feet in length, and are used for day fishing, rather than for long trips. In waters where the fishery is close to home, and where a minimum of gear and storage space is needed, these boats are often the ideal craft for the fisherman, due to their low initial price and maintenance cost.

The oyster garvey is a bit larger than most of the other outboard fishing craft, and is used in fairly rough waters, such as in many of New Jersey's coastal inlets. A sizeable fleet of these boats is found in Maryland's Chesapeake Bay oyster fisheries. Most are manned by only one person. The boats are about twenty-eight feet in length with a beam of seven feet.

(Sandra Buehler)

An unusual feature of this craft is the presence of two skegs (raised ridges) on the hull. They protect the boat if it runs aground by taking the abrasion themselves instead of allowing the hull's bottom to scrape the ground. Since this boat is often fished in shoal waters, the protection the skegs offer is quite important.

One of the most exciting ways there is to fish salmon is in a Pacific City, Oregon trolling dory. It is launched from the beach, and is rowed through the surf, using two pairs of oars. One set of oars, the set nearest the stern, is mainly for steering, and the man operating them faces forward. The other oarsman faces toward the rear of the boat and provides the force necessary to get the boat beyond the breakers.

After the boat makes its way past the surf, the little twelve-horsepower outboard is dropped into place, the centerboard is lowered, trolling poles are swung into place, and the boat begins to troll for salmon at speeds up to three or four knots.

These dories are made from Douglas fir plywood, are about twenty feet in length, and are usually of double-ender design, although some are built with wider "tombstone" sterns. The Pacific City dory derives its ancestry from the old Grand Banks dories that used to cod fish off Newfoundland in the late nineteenth century. It has retained that boat's flat sides and high bow, but it has been enlarged and greatly modified for higher speeds and more fish-carrying capacity.

MARINE REFRIGERATION

Refrigeration Methods

There are many and varied ways of keeping fish fresh until they are brought to port. The methods range from the very simple—like covering the catch with burlap bags and wetting the bags down with seawater—to the very complex multistage refrigeration systems found on the large tuna boats and factory trawlers. The method used is largely determined by the length of the trip. A day fisherman doesn't need anything more sophisticated than boxes of ice to keep his catch fresh, whereas a vessel leaving on a two-month trip must have a system capable of holding its catch at temperatures well below freezing.

Fish spoil in three ways. Bacteria inside and outside of the fish quickly cause the flesh to decay unless low temperatures are maintained. Enzymes, held at bay while the fish is alive, start to decompose parts of the fish as soon as it dies. Herring and sardines are especially susceptible to this type of decay. The third type of spoilage is caused by oxidation of fatty tissue that is exposed to the air. Oxidation results in a bitter and rancid taste, and a fishy, unpleasant smell. Fish with high fat content, like mackerel and herring, experience these oxidative changes quite readily. Freezing of the fish is more effective in controling bacterial decay than enzymatic or oxidative decomposition.

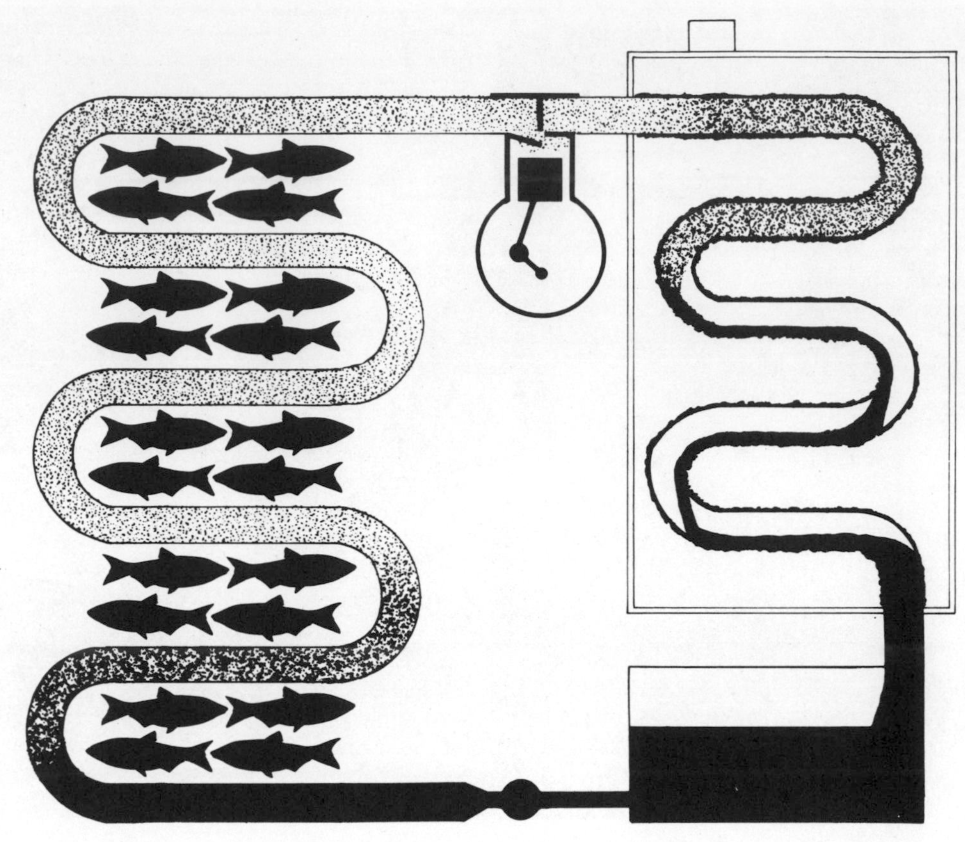

Marine Refrigeration. *(Ellen Storz)*

Icing the Catch

Keeping the catch fresh by holding it on ice is a simple and very good way of preserving it for short periods of time (ten days or less, usually). There is no complex refrigeration equipment to break down, and the initial cost of the vessel is quite a bit less than for a refrigerated boat (a seventy-five foot shrimp trawler with refrigeration costs about $20,000 more than one without). Icing is one of the fastest ways to cool the catch, which is important if its high quality is to be preserved. The fish are to a large extent cooled by the melting ice water that flows over them. Since freshwater ice always melts at 32 degrees Fahrenheit, keeping the catch on partially melted ice maintains it at a very constant temperature, which is another important factor for maintaining high quality. The melting ice water that flows over the fish also washes away some of the blood and bacteria.

The big disadvantage of icing is that it will not keep fish fresh for more than ten days. Often a very profitable trip has to be cut short because the boat has run out of ice, and four or five days of valuable fishing time are lost before the boat can return to port, unload its catch, get a new load of ice, and return to the fishing grounds. At the peak of the season in some southern ports, vessels have to wait two or three days before they can be re-iced.

In the Gulf of Mexico's shrimp fishery, shuttle ships cut down on lost fishing time by bringing loads of ice out to the boats, and taking their catch back to port. The shrimp are often damaged, though, because of the additional handling.

The method of icing depends on the type of sea life being caught; when shrimp or salmon are iced, care is taken to see that no two touch each other, for this would accelerate the spoilage rate. Every shrimp or salmon has a layer of ice between it and each of its neighbors. In addition, the salmon's belly and head cavities are filled with ice. With halibut, the body cavities are packed with ice, but none is placed between the fish, for this bruises the delicate flesh and lowers the market value.

Ice used aboard a commercial fishing vessel is manufactured in one of three ways. The first is by crushing large blocks or slabs of ice into irregular-sized lumps one-quarter to two inches in diameter. The large chunks sometimes make indentations in a fish's flesh, lowering its value. Flake ice, the second type used, is made by freezing water in thin layers on a smooth, chilled surface. It is then scraped off, or melted off by a quick, hot defrost. The third way of making ice is to freeze water inside of refrigerated tubes and then remove it by subjecting the tubes to a hot defrost. After this, the long, thin cylinders of ice are cut into short lengths.

All three methods produce ice of about equal quality for use on fishing boats. Storage life of the catch can be slightly increased by adding antibiotics to the ice. This method is rarely used, though. Sometimes ice is made out of frozen seawater instead of fresh water. Seawater freezes at a lower temperature, due to its salt content, and will thus keep the catch fresh longer.

Refrigerated Vessels

A vessel with a low-temperature refrigerated hold can stay at sea for as long as it likes, for deep-frozen fish can be kept fresh over a year. More fish can be taken, because valuable storage space is not taken up by ice. Refrigerated fish can be stored until market prices are at their maximum, whereas iced fish must be sold soon after they are caught, whether prices are good or not.

A comparison was made several years ago between two almost identical shrimp trawlers, differing only in that one had a refrigeration system aboard, and one did not. Both boats were seventy-two feet long, of steel construction, had three-thousand cubic foot fish holds, and were run by captains who were among the top 10 percent in the fleet, based on annual yield. The vessel using ice had less usable storage space and could not make as long trips, and thus landed only 70,000 pounds of shrimp during the season, as opposed to 84,000 pounds for the refrigerated vessel. Ten percent of the iced shrimp were damaged and brought only one-third the price of undamaged shrimp. The refrigerated vessel had only 1 percent of its catch damaged. It was paid a bonus, too, because the shore personnel didn't have to spend time icing it.

All in all, the refrigerated vessel grossed $25,000 more than the iced vessel that season. The refrigerated vessel had higher expenses, but even after they were paid off, it still was $10,000 ahead of the iced vessel.

Refrigeration Theory

A refrigeration system, whether it be a refrigerated hold, a quick-freezer, or a household refrigerator, employs a low-boiling-point liquid, called the refrigerant, that is used to cool metal plates, brine or air, which in turn cools the fish. Ammonia is often used as the refrigerant.

The refrigerant goes through several processes in order to freeze the fish. It first goes through an expansion valve, which lowers its pressure, thus lowering its temperature. Then it flows into an evaporator, where it cools down the metal plates, brine or air, that is used to chill the fish. The refrigerant has now warmed up, for it has taken some of the fish's heat. It has boiled away into a gas, for the boiling points of refrigerants are quite low (-20 degrees to -50 degrees Fahrenheit). After this, it is pumped through a compressor, which squeezes it into a more compact space, and pushes it toward the condenser. The condenser is a heat exchanger. It brings the hot refrigerant into contact with pipes of cool seawater, and allows the warmth from the refrigerant to flow into the water, thus cooling the refrigerant. The refrigerant turns back into a liquid as it loses its heat, after which it is pumped through the expansion valve again, and the cycle begins once more.

The quicker that fish are frozen after being caught, the longer they last. This is why vessels have quick-freezers as well as refrigerated holds. The time it takes to freeze a fish is dependent on how fast its heat can be transferred away from it and into the refrigerant. A freezer designed to chill fish quickly is usually one of three types: a cold-plate freezer, an air-blast freezer or an immersion freezer.

Cold-Plate Freezers

In a cold-plate freezer, the refrigerant is used to cool down a series of horizontal or vertical metal plates. The fish freeze quickly after they are placed in contact with the plates.

In a horizontal-plate freezer, fifteen or so plates are arranged one above the other, spaced about three inches apart (this distance is adjustable to accommodate different-sized fish). There may be several of these columns of plates in the freezer. The plates are opened to their maximum separation. The fish are placed between them, and the plates are brought closer together, until both sides of the fish are in contact with a plate. This reduces the freezing time. Horizontal-plate freezers are widely used for freezing blocks of filleted fish, and for whole, unfilleted fish frozen individually or in blocks.

Vertical-plate freezers are used for freezing whole bottomfish like cod, haddock or redfish. On the large factory trawler *Seafreeze Atlantic,* for instance, filleted bottomfish are frozen in one of three large horizontal-plate freezers that can handle 48 tons of fish a day. The fish that are not made into fillets are frozen in vertical-plate freezers that can handle 4½ tons a day, or in an air-blast freezer.

In the vertical-plate freezer, the fish are loaded through openings at the top of the space between each plate. They fall to the bottom under their own weight, forming a compact mass of fish that is in good contact with the plates. This allows the fish to be frozen quicker than in a horizontal-plate freezer.

Vertical-plate freezers produce a block of whole fish about three feet by two feet by four inches thick, weighing about ninety pounds. The block can be handled fairly roughly without breakage. Sometimes hydraulic systems lift these blocks out from between the plates, sometimes they are pulled out the side, and sometimes they are dropped out through a door at the bottom. They are then placed in a refrigerated hold leaving the freezer ready for more fish.

Freezer plates are flat, made of aluminum, and are typically ten inches by three feet by one inch thick. Several rectangular ducts run lengthwise through the plates, and it is through these ducts that the refrigerant flows. In order to free a block of fish from the plates after it has been frozen, a warm defrosting fluid is run through the ducts for a minute or so. This melts just enough ice so that the blocks can be easily removed.

Air-Blast Freezer

An air-blast freezer cools by circulating chilled air among the fish, which are placed on racks inside the freezer. It produces fish of higher quality than those of the plate freezer. It produces fish of higher quality than those of the plate freezer, which sometimes deforms the fish due to uneven cooling. It takes more time and more energy to air-freeze fish, though, because moving air does not carry the heat away as fast as a metal plate does. Metal is an extremely good conductor of heat, whereas air is a poor one.

The air moves at velocities of fifteen to twenty-five feet per second. It is chilled by being blown through a heat exchanger composed of rows of refrigerant-filled tubes. The tubes have fins on them, which increase their surface area and heat-exchanging properties.

Some freezers let the air circulate by natural convection instead of by moving it with a fan. These freezers tend to chill the fish slowly and unevenly, for the warmer air will sit at the top of the freezer, and the cooler air will remain at the bottom.

Immersion Freezers

Fish are sometimes frozen by immersion in cold brine. By adding salt to water, a brine solution can be made whose freezing point is well below 0 degrees Fahrenheit. If this solution is chilled to a temperature above its freezing point, but below the freezing point of the fish (which is about 28 degrees), fish placed within it will quickly freeze, even though the brine will remain liquid. It is much more efficient than the air-blast freezer, for water has the ability to absorb a lot more heat than the same volume of air (this property is known as heat capacity). Immersion freezing is used quite often in the Pacific tuna fishery, and also on some large halibut and salmon boats.

Fish immersed in brine absorb some of the salt, which changes their flavor. This is a disadvantage if the fish are bound for the fresh-frozen fish market, but not that serious if the fish are to be canned.

Brine sloshing back and forth can make a vessel unstable, especially in a rough sea. In order to guard against this, holds are divided into groups of fairly small tanks that minimize the sloshing.

The Refrigerated Hold

After the fish are removed from the freezer, they are placed in a chilled hold, to remain until the vessel reaches port. Holds are kept at temperatures as low as 20 degrees below zero.

Whole fish come out of the quick freezer singly or in blocks, and are stowed in the hold without being wrapped. Fillets and other processed fish are wrapped or packed in cartons before stowing.

The most popular method of cooling the hold is to have grids of galvanized steel piping, with refrigerant running through them, mounted on the walls.

Some holds are cooled by a stream of forced air circulating through them. The apparatus for this method takes up less space and is less expensive to install than the grid systems. It does not cool as evenly, though, and uses more power. Other holds have a sealed air space in between the insulation and the inner lining of the hold. Cooled air is circulated by a fan through this air space.

How Different Species of Fish Are Handled—Tuna

On the smaller boats, the tuna are laid on deck and covered with wet burlap bags until the heat they give off as they die is dissipated. Then they are frozen "in the round" (they are not gutted or filleted). They are laid out in the refrigerated hold until the bottom of the hold is covered. By this time, the first fish laid out are frozen on the outside. Then another layer is started on top of the first, and so on until the end of the trip.

On the large tuna boats, the catch is put into seawater-filled wells held at 30 degrees. After the well is filled (this generally takes eight days or less for a fifty-ton well), enough salt is added to lower the water's freezing point to about 10 degrees. This brine solution is then chilled to a bit above its freezing point. In a day or so, the fish are frozen and the brine is pumped out. The fish are kept frozen until port is reached by a series of refrigerated pipe grids in the walls of the well.

Salmon

Salmon that is to be canned is sometimes frozen, rather than stowed on ice. Before freezing, it is first gutted, washed, and thoroughly bled. Any blood on the flesh discolors it and reduces its market value. After this, the fish are frozen in an air-blast freezer. When they are removed, each fish is "glazed" by dipping it into a tub of seawater. The thin layer of ice that forms around each salmon locks the moisture in, thus lengthening the time the catch will stay good. After this, they are put in bins in a refrigerated hold. The bins are lined with large plastic bags to further reduce dehydration of the fish.

Shrimp

Shrimp are handled in several different ways. In the first way, the shrimp are put into fifty-pound open-mesh bags and lowered into an immersion tank of chilled brine. It takes only fifteen to thirty minutes for them to freeze. After removal from the tank, they are glazed, using a corn syrup and brine solution, and are allowed to drain for a few seconds. Lastly, they are stowed in a refrigerated hold.

Some boats use an air-blast freezer to cool their shrimp. The shrimp are bagged and lowered into the hold. The bags are laid on racks and flattened as much as possible. They are left in the refrigerated air blast for six to eight hours, after which they are glazed and stored in the hold.

In refrigerated seawater (RSW) systems, the shrimp are placed in a tank filled with brine at about 30 degrees, just above the freezing point of the shrimp. They can be kept for up to sixteen days in this manner.

Lobsters

Although the common way of preserving lobsters is to keep them in a live tank filled with seawater, some boats freeze their catch. After emptying a trap, the fishermen remove the tails and place them in plastic or open-mesh

Loran receiver. *(Sperry Marine Systems)*

bags, put them into a refrigerated air blast for a day, glaze them and store them in a refrigerated hold.

Menhaden

Pogies are stored for up to a week in 30-degree RSW tanks. Since they are used for cat food and reduction, not much attention is paid to sanitation or keeping a constant storage temperature. Unloading is accomplished by pumps or vacuum systems.

MARINE ELECTRONICS

Electronics devices on fishing boats serve two purposes—to help in navigation, and to find fish. It used to be that fishing boats had to find their way by dead reckoning, celestial navigation and perhaps with the aid of a compass. Today, sophisticated navigational aids such as loran, radar and radio direction finders are becoming increasingly common on small boats as well as large ones. Radios and fish-finding depth sounders are also becoming more common.

Loran: Boats use loran to locate their position on the ocean. A loran receiver is a device that measures the time difference between the arrivals of two loran radio signals from two known locations. The closer the boat is to one of the stations, the greater the time difference. By referring the time difference to a loran chart or table, a loran line-of-position (LOP) is immediately found. Intersection of two or three LOPs gives the ship's position. Captains often record the loran readings of the places where they set their traps. This enables them to find the traps again, even though they may be fifty miles offshore.

Radar: The word radar comes from the words "radio detection and ranging." Radar is used for navigating in coastal areas and narrow channels when visibility is poor. It is also used for avoiding collisions with other boats. Radio waves are transmitted from the radar set's antenna. When

Water temperature indicator. *(Danforth)*

Autopilot. *(Danforth)*

they hit a rock or another boat, they bounce off and are picked up by the antenna when they return. The difference between the time the signal was sent out and the time it was received gives the distance of the object, and the direction the antenna was facing gives the object's direction.

Radio Direction Finder (RDF): RDFs give, as the name implies, the direction that a radio signal is being transmitted from. The RDF antenna is turned by hand to various posi-

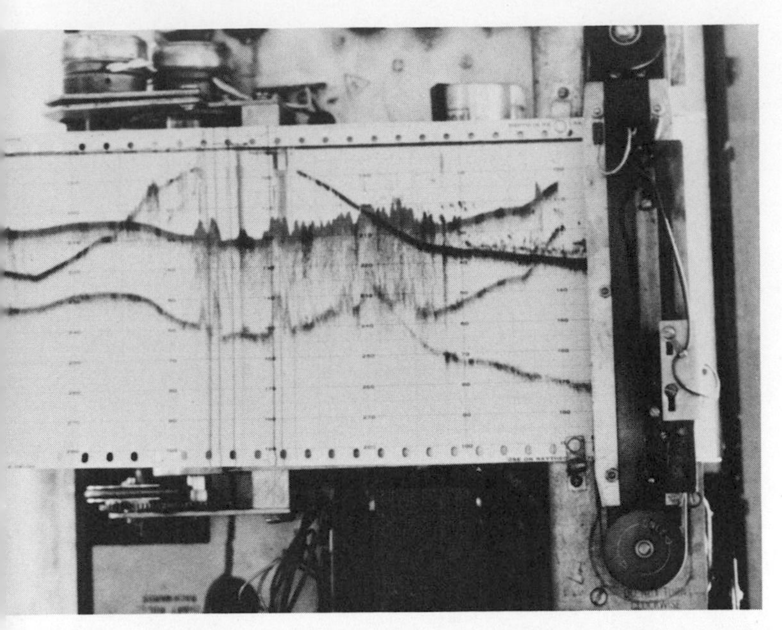

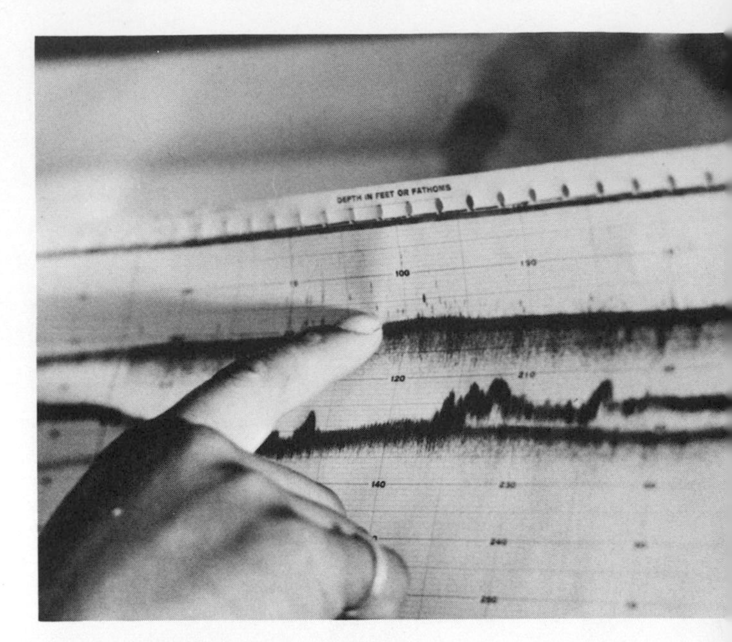

Sensitive recording depthfinder. Small peaks indicate individual fish. *(Rochelle Smith)*

Depth finder. *(Danforth)*

tions. When the signal is weakest, the antenna is facing directly away from where the signal is coming from.

RDFs are most commonly used as homing devices by locating the direction of a signal from some port. They are also used to locate other boats. If a skipper announces over his radio that he is "in the fish," other skippers will home in on his radio signal.

Some of the newer boats have ADFs (automatic direction finders) that are similar in principle to RDFs, but locate the direction of a signal without having to be turned by hand.

Radios: Citizen's Band radios, nicknamed Mickey Mouses, are very common on fishing boats. Many automobiles and trucks use them, though, and the air waves are frequently too cluttered with conversations for fishermen to get through to other boats. Fishermen are being forced to go to other types of radios that transmit at different frequencies, such as Single Sideband radios, AMs or VHFs (Very High Frequency radios). A Single Sideband has a range of hundreds of miles and can often pick up far-away weather warnings that give an offshore boat plenty of time to reach port before a storm hits.

Water Temperature Indicators: The temperature indicator's sensor is located on the outside of the hull and the indicator's readout is sometimes in the wheelhouse and sometimes outside. Boats fishing for temperature-sensitive fish use water temperature indicators to help find the fish. For instance, albacore are most likely to be caught in waters with temperatures between 60 and 68 degrees Fahrenheit.

Autopilot: An autopilot consists of a compass and control assembly. A sensitive magnetometer in the compass detects when the boat veers away from the desired heading and sends an error signal to the control assembly. The amplifier in the control assembly generates a signal that energizes either an electric steering engine if the boat has mechanical steering, or control valves if the boat has hydraulic steering. These devices turn the rudder until the error in direction is corrected.

Depth Finders (also called *Echo Sounders* or *Flashers*): The concept of a depth finder is similar to that of radar, except that radar uses radio waves and depth finders use sound pulses (the pulses are usually beyond the range of

Fisherman with CB radio. *(Maine Department of Marine Resources)*

human hearing). They have long been used to avoid underwater obstructions and shoal waters and are now being employed as fish finders as well. The sound pulses they emit bounce off rocks and schools of fish. Good depth sounders can even pick up the echoes from individual fish.

Sounders consist of an indicator or recorder and a transducer. The indicator or recorder gives a reading of depth on a dial or digital display, or charts a permanent record on a graph. The readout tells the distance to the bottom or to a school of fish. The second part of the sounder, the transducer, is mounted on the hull and is connected by wires to the indicator. The transducer emits a sound pulse and picks up reflections of the pulse.

Coded Radio Messages

"Hey, Al, you listening?" the captain calls softly over the radio. "Not even a piss-ant in the last lick. Barely worth trying." A message like this might mean that the fishing is no good, or it might mean that the captain is up to his ears in fish, and Al had better get his tail over there if he wants any.

Fishing fleets are often broken down into little subgroups of four or five boats that fish with each other and try to help each other out whenever they can. A skipper knows the other men in his group quite well, and a small inflection in any of their voices can mean a great deal. "This is Billy-Boy, and I'm about three miles north of the bar. Nothing much over *here*." A broadcast like that might send his friends running to join him. Billy-Boy doesn't want too many boats around him, though, or he is liable to be crowded out of his own area. He wants to give just enough information to tell his friends what is happening.

One old Portuguese captain used to come on the air now and then with a message like, "Jorge, you might try a pass over here." All who knew the captain knew he had found one hell of a school of fish. Any skipper lucky enough to tune in on the conversation would drop what he was doing, get a fix on the old captain's boat with his radio direction finder, and make it over there as fast as he could.

Sometimes a skipper will yell "Switch!" and he and his friends will all flip their radios to a different, prearranged frequency. This tactic is supposed to befuddle anyone listening in on their conversation. But many boats are now equipped with "scanners" that can quickly locate the frequency a boat is broadcasting on.

A few of the boats even have "scramblers," like the FBI uses when they don't want their telephone calls intercepted. The device scrambles and garbles a man's voice so that it's unintelligible. Only the man's friends will have the proper descrambler on their radios.

Every now and then, a fisherman will find a school of fish big enough for the whole fleet, and he'll announce this fact over the air waves. He had better be sure the school is big, though, because his message could bring a hundred boats in a matter of minutes.

Fishing Gear

Trawls

A trawl is a cone-shaped net that is dragged behind a fishing vessel, scooping up fish as it goes. The net is generally weighted down and towed along the bottom, although midwater trawls that float a considerable distance above the bottom are also used.

There are various ways of keeping the mouth of the trawl open. The most common is through the use of otter boards (also called trawl doors or just plain doors). The otter boards are massive slabs of wood or steel. One is attached, using cables, to each side of the trawl's mouth. Other cables run from the doors to the boat. The cables are attached to the doors in such a way that, as the boat tows them, the doors sheer outward, holding the trawl's mouth wide open. The bottom of the trawl's mouth is weighted with chains, to keep it on the ocean floor, and the top of the mouth is rigged with floats, so that the mouth stays open in a vertical direction.

The mouth of the net swallows up all fish in its path. The fish are eventually pushed down the length of the net into the narrow rear part, called the cod end. The trawl is towed for twenty minutes to four hours, depending on how fast it fills up with fish. It is then hauled out of the water, and its contents are dumped on the deck to be sorted. The trawl is immediately returned to the water, so that no fishing time is wasted. Otter boards on a trawler stand about ten feet high, four feet wide and four inches thick. They weigh about 1500 pounds apiece. Nets using otter boards are called otter trawls.

Cod trawlers use a "3:1 scope" of cable length versus depth of bottom. In other words, if a vessel is dragging in water 100 fathoms deep, there will be 300 fathoms of cable between it and the otter boards. This 3:1 ratio allows the net to tend bottom (drag smoothly along the ocean's floor). A shorter cable would jerk the net off the bottom, and many groundfish would be missed.

A trawl can be adjusted to fish on different types of bottoms by adjusting the length of the ground cables, the cables that connect the otter boards to the net itself. The longer the cables, the better the net stays on the bottom. It is fine to use long cables if the bottom is smooth and flat, but if it is uneven and rocky, the net should be rigged so that it can bounce over obstructions, rather than snag on them and tear. This is done by shortening the ground cables.

The forerunner of the modern otter trawl was the beam trawl, so called because its mouth was held open by a rigid beam. The large and heavy beam was dangerous to handle in a rough sea. It was an appropriate type of trawl for the old sailing vessels, though, for the beam kept the mouth open, no matter what speed the boat sailed at. An otter trawl would not have worked as well, for if the wind conditions are gusty and changeable, a boat's speed would constantly be changing, which would alter the tension on the cables. A slackening of the cables would close up the trawl mouth.

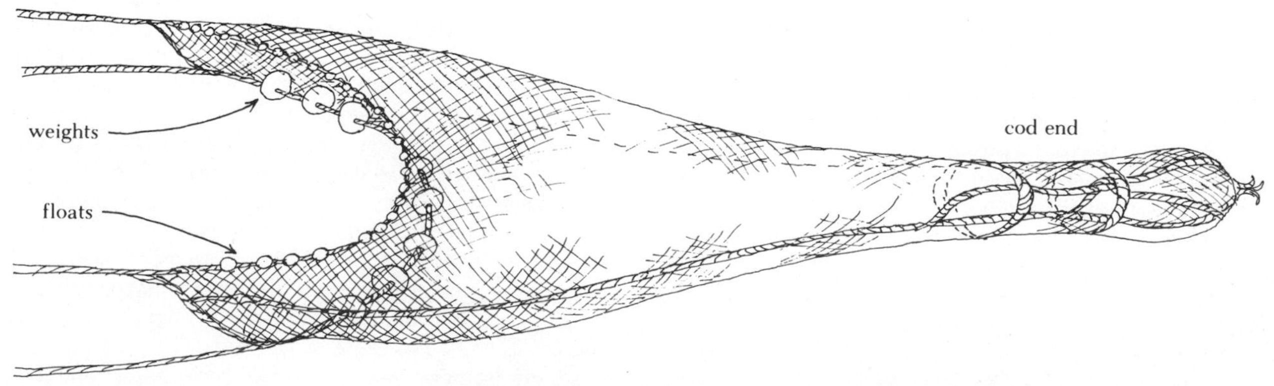

Otter trawl. *(John Douglas Moran)*

Otter trawls came into use after the advent of steam power. A power boat is able to trawl at a constant speed, no matter what the wind conditions. The floats fastened to the top of an otter trawl's mouth enable the mouth to open in a vertical direction more than a beam trawl's mouth. Due to this feature, when otter trawls were introduced, catches of roundfish (cod, hake, haddock, pollock, etc.) greatly increased. This is because roundfish spend their time slightly above the bottom, and were often a bit too high for the beam trawls to scoop up.

The mouths of otter trawls can be made much larger than those of beam trawls, and thus can catch more fish. A beam trawl's mouth is limited in size by the length of the beam and by the height of the metal shoes that keep the mouth open in the vertical direction. The length of the beam is in turn limited by the size of the boat. An otter trawl has no such limitations, for its mouth is held open by floats and by the sheering outward of the doors. Beam trawls are still used on some boats (for instance, some of the drag boats in the Boston fleet and some shrimp trawlers on San Francisco Bay still use them), but they are vastly outnumbered by those using otter trawls.

The life of a fisherman has changed greatly since trawling replaced dory fishing and handlining from schooners. The men on the dories had to rely completely on their own skill or luck to catch any fish. On a trawler, however, the crew works as a team, with each man being assigned a fairly specialized task.

Purse Seines

A purse seine is a large net used to encircle a whole school of fish at one time. It gets its name because after the school is surrounded, the bottom of the net is pursed, or drawn shut, preventing the fish from escaping by diving. Over half of the total United States catch is landed in purse seines. The world's big reduction fisheries (where the catch is made into fertilizer and fish meal, rather than used for human consumption) all depend on purse seining.

The major North American purse seine fisheries are: The East Coast and Gulf of Mexico menhaden fishery (menhaden accounts for most of the North American purse seine catch, and is used for reduction only), and the West Coast and Alaskan tuna and salmon fisheries. Although the total annual tonnage of menhaden far exceeds that of salmon or tuna, the salmon and tuna catches are each worth more money than the menhaden catch, for the fish are worth much more per pound than menhaden. Each fishery has its own methods of purse seining and its own special boats. The menhaden fisheries use the old purse-boat method of operating the seine, and the salmon and tuna fisheries each use variations of the more modern catcher boat and skiff method.

In the East Coast menhaden fisheries, two purse boats (32- to 36-foot steel or aluminum craft equipped with power blocks for hauling the seine) are carried to the fishing grounds on board a large steamer. When the steamer reaches its destination and the purse boats are lowered into the water, each boat contains half of the large purse seine. The boats run side by side until a school of menhaden are located. The boats approach the school and then split up and go in opposite directions, playing out the net as they go, surrounding the school with it, and finally pursing the net (drawing the bottom shut).

After the pursing operation, the crews of both boats begin to draw the huge net aboard. This operation, called drying up the seine, used to be done by hand, and was backbreaking, time-consuming work. Now it is done with hydraulic power blocks which cut down greatly on the number of crewmen needed.

The net is hauled until the part of the net that is still in the water is thickly populated with menhaden. At this time, the steamer comes by and brails the fish aboard (transfers them onto the steamer by means of a smaller net that is repeatedly dipped into the purse seine), or sucks them up with a large hose and pump.

After the fishing is done for the day, the boats are hauled aboard the steamer. In rough seas, launching and hauling the purse boats can be dangerous, and can use up valuable fishing time. Also, the steamer must be fairly large in order to carry the two boats. This means that this method can only be used where there are enough fish caught to maintain the steamer and two purse boats, and still make a profit. It probably would not be profitable to use it in the West Coast salmon fisheries, for the schools of salmon are much smaller than menhaden schools.

The menhaden fishery is dangerous to work on in rough weather, for the small purse boats can easily be swamped. Also, the size of the net is limited by the size of the purse boats that carry it. These boats have the advantage, though, of being speedy and very maneuverable, enabling them to quickly surround a school of fish before they dive and escape.

The Catcher Boat and Skiff System

In West Coast fisheries, the purse seine is carried on board the catcher boat, which can be a 58-foot Alaskan salmon seiner or a 295-foot California tuna superseiner. When a school of fish is located, one end of the purse seine is attached to the skiff, and the skiff is lowered into the water. The other end of the net is attached to the catcher vessel. The fish are surrounded and the net is pursed and pulled aboard the catcher vessel using a power block. The fish are usually brailed aboard the vessel rather than being sucked up in a hose, as are menhaden, for it is important that tuna and salmon catches be as little damaged as possible. Since menhaden eventually become fertilizer or fish meal, it is not very important if they are crushed and torn as they are sucked up in the hose.

There are advantages of the catcher-boat-and-skiff system over the purse boat system. Most of the crew remain aboard the catcher boat, which is bigger and more seaworthy than menhaden purse boats. Thus, in a rough sea most of the crew are exposed to less danger. Also, since the seiner is much bigger than the purse boats, it can carry a larger net, increasing the potential catch. Tuna seines are sometimes a mile long.

Salmon seiners use a smaller, shallower net than menhaden or tuna boats, for salmon schools are smaller,

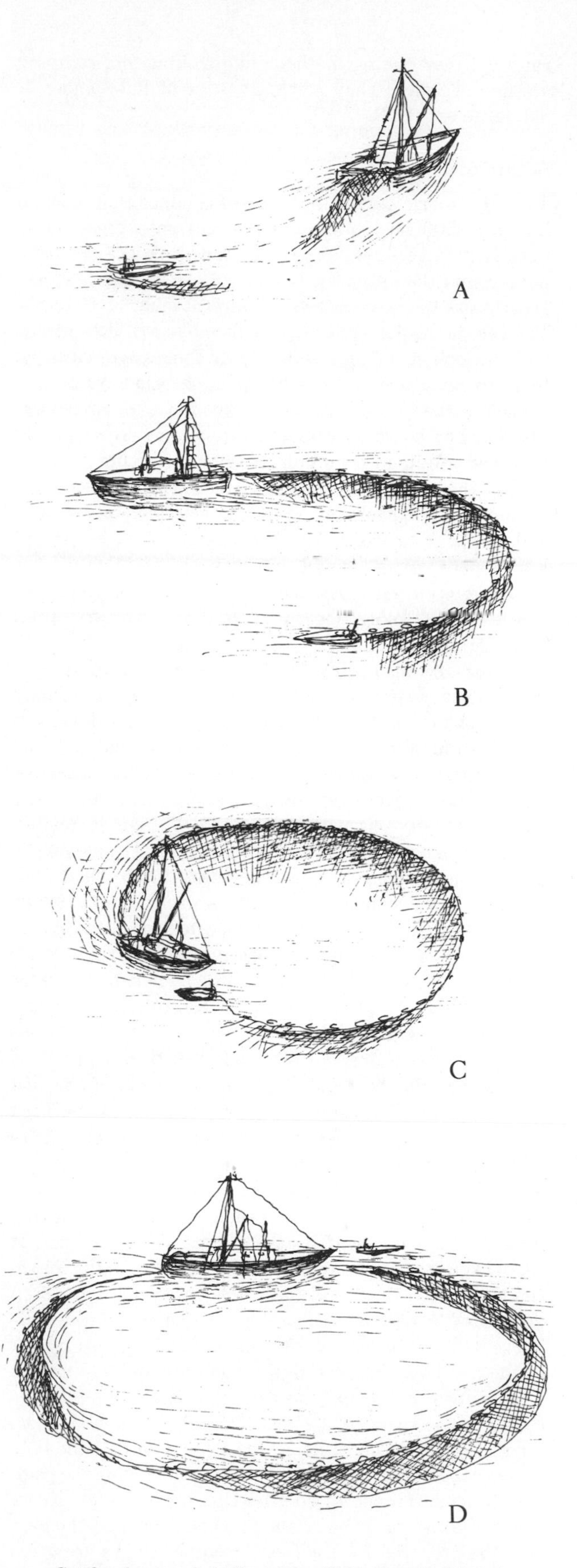

Catcher boat and skiff system. *(John Douglas Moran)*

travel close to the surface and are caught in fairly shallow water. Some salmon seiners wind the net onto a drum while others haul it with a power block and pile it on the deck as it comes through the power block. Unreeling the purse seine from the drum can be hazardous, for sometimes it catches on itself and backlashes. The skiff pulls the net away from the catcher boat as it is being unreeled. A backlash will jerk and sometimes capsize the small skiff, which can be a serious accident in near-freezing Alaskan waters. One skiffman tells of how just this thing happened to him. When he felt his skiff starting to roll, he immediately leapt clear of it, for he did not want to be trapped underneath. The next thing he did was to rid himself of his heavy hip boots, his rain parka and his jacket, so he would not be weighted down and drown (he always wore parkas that zipped down the front instead of slipover parkas, for slipovers are much harder to take off in the water). Then he made his way to the capsized skiff, which still had one end of the purse seine attached to it. The captain pulled him and the skiff up to the catcher boat by reeling the seine onto the drum.

The Power Block

The biggest change in purse seining in recent years was brought about by Mario Puretic's invention of the power block. It used to take twenty-four men to haul a menhaden seine by hand. The power block cut this number to twelve, and reduced the time it takes to haul and pile the net.

The power block is basically a large pulley with a deep V-groove in it, into which the net fits. As the block turns, the friction between the net, jammed into the V-groove, and the block itself is sufficient to haul the net. The net goes up and over the block and then plunges to the deck below, where deckhands wait to stack it neatly as it comes down. Most modern power blocks are hydraulically driven, although the first models were mechanically turned.

Puretic was a Yugoslav fisherman in the Southern California sardine, mackerel and anchovy fisheries. In 1954 he developed a prototype of the block, driven with ropes and attached to a boom. At first, he could not convince anybody to use it. Then, in 1955, Marine Construction and Design Company (Marco) in Seattle heard about it and invited him to come up and show them how it worked. He drove to Seattle with one in the trunk of his car. They agreed to produce it. In a few years, mostly because of the power block, purse seining replaced pole and line fishing as the predominant method of the American Tuna Fleet. As of June 1975, Marco had sold 11,000 power blocks in about forty countries.

Gillnets

Gillnets are nets whose mesh size is just big enough so that the fish they are meant to catch can slip partway through the mesh. When they can go no further, they try to back out, only to find that their gill covers have caught in the webbing, trapping them. The gillnets are walls of netting six to ten feet high, on the average, and several hundred or

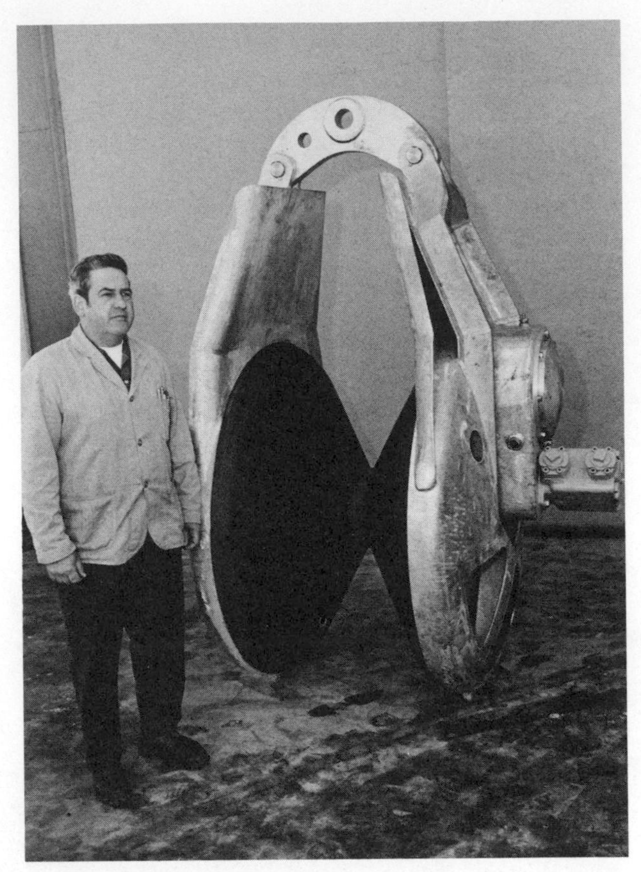

Puretic power block. *(Bob Carver)*

more feet long. They are weighted on the bottom to keep them on the seafloor, and have floats on the top to keep them vertical. They are placed so that schools of fish will swim into them and get entangled. They are often stretched across rivers to intercept schools of salmon or smelt on the way to their spawning grounds. The ends of some nets are attached to buoys. Some nets are allowed to drift with the tide, and others are anchored in place. Crude anchors are often made by filling a fifty-five gallon oil drum with rocks. Buoys are at times nothing more than inner tubes with kerosene lanterns mounted on them to help locate them.

There are two methods of removing fish from the net. If the gillnet is anchored in shallow water, some fishermen run their boats parallel to the net, quite close to it, and haul up sections of the net. They pick the fish out of them, drop the section of net back into the water, and move the boat along to a new section. Others reel the net onto a hydraulically powered drum mounted on the boat, picking out the fish as the net comes aboard. The men must be skilled in order to quickly disentangle the fish from the net as it comes aboard. Dogfish and spiny rockfish are especially hard to remove.

An advantage of gillnetting over some other types of fishing is that it does not kill immature, undersized fish, for they are able to swim right through the webbing of the net. It is important that enough of these young fish be allowed to mature and spawn if the fishery is not to be ruined. Other fishing methods like trolling and setlining are non-selective. They catch all types of fish, young or old, large or small.

Setlining

Take the snarls out of 40,000 feet of line, chop enough bait for 2500 hooks, squeeze the bait onto those hooks until your fingers are cut so badly they look like raw hamburger, and then play the line out of the baskets they are coiled in as the boat barrels forward at eight or so knots. You see the hooks whip by your face like evil little darts, and you look at the razor-sharp knife that is kept close by, there to cut you free if one of the hooks snags you.

Each of the tasks mentioned above must be carried out every day on board a setliner, sometimes by crews of only two men. Setline crews don't spend much time in their bunks. They get to rest for a few hours when the setline is sitting on the bottom, after which they have to reel it in and get back to work. The line passes through a set of knives as it comes aboard, which strips the fish off the hooks. A crewman coils the incoming line as neatly as possible as it comes aboard, so that he won't be up all night untangling snarls.

Setlines are also called longlines, and for good reason, for some are twenty or more miles long. They are usually laid on the ocean floor to catch groundfish, although in swordfish longlining, the line is buoyed so that it floats only a little way below the surface. At regular intervals on the setline (every few feet if fishing black cod, every twenty or thirty feet with halibut) short pieces of smaller-diameter line, called gangions (pronounced "gan-yuns"), are attached. A baited hook is fastened to the end of each gangion. At each end of the setline is a warp line that runs to a buoy on the surface. The boat locates one of the buoys, pulls alongside it, brings it aboard and slips the line attached to it around the hydraulic line hauler. If the water is clear, a long parade of hooked fish can be seen approaching the surface as the line is reeled in. As the line comes aboard, it passes over a roller and then through a set of knives that rip the fish off the hooks. Some of the fish come up mere heads and skin, their insides having been eaten by slime eels. These creatures enter through a fish's gills after it is trapped on the hook, and methodically chomp away at its innards. Other fish come out of the water minus everything but their heads, their bodies having been devoured by dogfish, the plague of the setliner. It takes experience to know just how long to leave the line in the water. Too long, and the whole catch will have been decimated by predators. Too short a time, and most of the hooks will still be empty. The line has to be pulled up out of the depths at just the right angle, too, or the line will get strained and snap. The hauling angle is adjusted by changing the speed of the boat. Many boats have dual engine and steering controls near the stern, so that the fisherman can haul or set a line while he is managing the boat.

There is still much to do after the line is hauled. Every one of the fish must be cleaned and washed, and the gear must be prepared for the next morning. Hooks are often bent when the line is pulled through the knives. These

(Port of Los Angeles)

(Port of Los Angeles)

hooks have to be straightened or replaced. Worn gangions must also be replaced. The miles of line must be coiled neatly and placed in baskets (called "tubs") with the hooks canted just right so that they will not foul when the line is paid out.

Most of the longlining in the San Francisco area is for sablefish, better known as black cod. There is a sizable halibut setline fleet working out of Washington and Canada, composed mostly of men of Scandinavian backgrounds. Setlining is the only legal way to catch halibut. Trawling for the fish is not allowed, and trawlers must throw back all halibut that happen to come up in their nets.

Longlining for swordfish is done off the New England coast. Most groundfish on the East Coast are caught by trawlers, although some are still caught on longlines. In the old days, all of the rich cod and haddock grounds, like Georges Bank and the Grand Banks, were fished with setlines. A fleet of dories would fish as many setlines as there were dories, and return each night to the mother schooner with their catch. Portugal still sends a fleet of dory schooners to the Grand Banks each year.

Japan longlines much more extensively than the United States or Canada. They send a fleet of huge factory boats to the Gulf of Mexico each year for tuna. The lines they set are sixty-five nautical miles long! The boats have 150-man crews and catch ten million pounds of fish during a nine-month season. The Japanese have their Gulf of Mexico tuna operations down so well that they can catch, process, can and ship the tuna back to America far cheaper than Americans can catch and process this Gulf of Mexico tuna. Because of this, the Japanese have a virtual monopoly on gulf tuna.

Other fish come up on the Japanese longlines, too, but they throw none of it away. Marlin is made into a type of sausage that is considered a delicacy, and is eaten with horseradish and soy sauce. Swordfish and shark are also very popular.

Trolling Gear

When trolling, a boat dangles a number of lines behind it as it runs through the water. The lines each have one to several hooks on them, and fish are attracted either by bait or by lures. When a fish strikes one of the hooks, the line is hauled either by hand or by a hydraulically driven spool (some are mechanically or electrically driven).

Trolling vessels are easy to recognize by the two tall poles located amidships, on either side of the mast. When fishing, these poles are lowered to about a forty-five degree angle, and fishing lines are attached to them. The poles are roughly the length of the boat, and are made of cedar, fir, hemlock, spruce, aluminum or steel. If wood poles are used, cedar is best because it resists decay. Many of the poles are made by the fishermen themselves by cutting down and roughly finishing a tree of the right height and diameter. It is best if the wood in a new pole is allowed to dry, but often, if a pole breaks, the fisherman will pull into a harbor, cut down a tree and rig it into a trolling pole right on the spot.

Salmon trolling gear is fairly complex. The stainless steel fishing lines are wound on spools called gurdies that are usually turned hydraulically, although some are mechanically or electrically driven. Small boats, like those of the Pacific City, Oregon dory fleet sometimes have hand gurdies. There is a gurdy for each line. Small boats might have only two gurdies; larger boats, up to eight. The fishing lines run from the gurdies through a set of blocks to the trolling poles and then into the water. The fishing lines are attached to the poles by means of short tag lines, at the ends of which are the clothespins, through which the fishing lines slide. Bells are mounted at the spots where the tag lines meet the trolling pole, so the fishermen will know when a fish has taken a hook.

Different lines are attached at different spots to the trolling poles, so that one line will not become entangled with the others. The lines are of different lengths, and are weighted so that they will sink to different depths. This also helps prevent them from fouling with each other, as well as enabling the boat to fish at several different depths simultaneously.

Attached to each fishing line are several spreads. Each spread consists of a short line, at the end of which is a hook. The hooks are sometimes baited (herring and squid are often used) and at other times, "hardware" is used to lure the fish to the hook. Which method is used depends on the skipper's preference. Salmon trolling is an art that takes years to learn, and each skipper has his own ways of doing things. Some start out with lures and "hoochies" in the beginning of the season and switch to bait later on. Other fishermen stick with one or the other.

The types of lures used are many and varied. Flashers are flat, shiny pieces of metal that look like small fish to the salmon. These flashers are made so that they will go through the water on edge, rather than lying flat, for if they lie flat, they will look like dead fish, and the salmon won't be interested. Hoochies are rubber and plastic lures that resemble little squid. They get their name because, when being pulled through the water, they jiggle around "like a hootchy-kootchy dancer." Also, the plastic streamers attached to the lures look sort of like a woman's skirt.

At the end of the fishing line is a short nylon or rawhide breaker line that is purposely made a lot weaker than the stainless steel fishing line. At the end of the breaker line is a lead cannonball sinker. The purpose of the breaker line is to give way if the sinker snags on something on the bottom, thus preventing the main line from breaking. If the main line broke, quite a lot of expensive gear could be lost. The depth that the fishing line sinks to is controlled by adjusting the line's length and the weight of the sinker. When trolling, the boat runs at one to three knots. If a very light sinker is used, the line will stream backward from the trolling pole, staying almost on the surface of the water. The heavier the sinker used, the closer to vertical the line will hang from the trolling pole. The skipper tries to have lines at all the depths where the fish are likely to be swimming. The inside lines are usually the deepest, and carry forty- to sixty-five-pound sinkers. The middle lines carry the least weight—twenty to thirty-five pounds. They are the longest and shallowest of the lines.

The center of activity in the boat is the trolling pit, a several-foot-deep trench near the stern of the boat where the fishermen stay while trolling. The gurdies are controlled from there, the fish are gaffed out of the water from there, and the boat can usually be controlled from there, too, so that the men don't have to leave the pit. In the pit are knives, hoochies, gaffs, clubs, flashers, bait, line and several pairs of gloves, usually with a rough finish to them, to enable the slippery fish to be gripped firmly.

When the bell at the end of the tagline rings, letting the fisherman know that a fish has struck the hook, the fisherman starts the proper gurdy rotating and reels in the fish (some let the fish "soak" for a while before reeling it in, hoping that its thrashing will attract other fish). When the fish breaks surface, the fisherman reaches over the side with a gaff and snags it in one of its gills. If he snags it elsewhere, the mark will show and lower the price of the fish. He hauls it aboard and kills it by hitting it over the head with the gaff handle or a club. Then he removes the fish from the hook and, as soon as he can, guts and cleans it, and throws it into a tub of seawater, to await icing later in the day.

Albacore Trolling

Albacore trollers barrel along at three to four times the speed of salmon trollers—up to eight knots. They fish far offshore and are thus quite vulnerable to rough weather. The albacore boats are often too far offshore to reach port before a gale hits and passes them, so they just stay where they are and try to weather it out. In 1962 an unexpected typhoon hit the West Coast and caught a number of albacore boats in it and about twenty men died.

Albacore trolling gear is quite a bit simpler than salmon gear. There are no gurdies, blocks or lead sinkers. The fishing lines run from the trolling poles into the water, and have only one hook attached to the end. Lures similar to the salmon troller's hoochy are used. Lines called the haul lines are attached to the fishing lines at the points where they enter the water. The haul lines run from the fishing lines to the boat. On most boats when a fish strikes a hook, the fisherman pulls in the haul line by hand until he can reach the fishing line, and then pulls that in by hand. Some boats are equipped with automatic line pullers that save the fisherman this work. Up to eight lines can be attached to each trolling pole, twice as many as with salmon. Five or six lines on each pole are more common, though. Several more lines might dangle from the boat's stern. The lines are nylon, and are not nearly as strong as the 900-pound test stainless steel salmon lines. This is because the lines do not have to carry heavy cannonball sinkers and up to a dozen hooks on them.

Pots

Pots are basically cages that are equipped with funnels or trap doors that allow fish or crustaceans to enter, but not to leave. The words pot and trap are used fairly interchangeably when referring to these devices. Pots are made in scores of different shapes and sizes. New England lobstermen use half-round traps shaped like a cylinder sliced in half lengthwise. Alaskan king crabs are caught in large rectangular traps that weigh up to 800 pounds. The traps are quite strong, in order to withstand Alaska's rough wintertime seas. Dungeness crabs are taken in squat, round traps. Most West Coast traps are made of metal, although West Coast shrimpers still use wooden pots. The frame of a metal pot is usually made of steel bars welded together. The meshing is stainless steel screen.

In New England, both wooden and metal traps are common. The wooden traps are made of cedar slats, or lars, nailed onto oaken frames. These traps have to be weighted with bricks or rocks to keep them on the bottom. Materials for an inshore trap cost about ten dollars. The traps are often made by the lobstermen themselves.

Traps are fished singly, doubly, or in large groups. The singles are attached by a line to a buoy on the surface. In the doubles method, two traps are attached to each buoy. Large groups of pots are fished in much the same way as setline gear. In fact, setline gear is sometimes used in trap fishing with the hooks being replaced by traps.

Traps employ different methods to allow the young, undersized fish or crustaceans to escape. Dungeness crab pots have escape holes a few inches in diameter built into them. On some wooden lobster traps, the slats are spaced far enough apart to allow shorts to escape. In other traps, the shorts are simply removed and thrown back when the trap is hauled into the boat.

A serious problem that can potentially ruin a fishery is what to do about ghost pots. Pots that become separated from their buoys can continue to fish for years after they have been lost. A fish will enter and eventually die, and serve as bait for other fish to enter the trap. Thousands of traps are lost each year, and they represent a considerable danger to the fisheries. Newer traps are being equipped with destruct panels. Some of the webbing is made of a fiber that will disintegrate after a few weeks or months, allowing all fish that enter the trap to escape.

Ethnic Fishing Communities

North America has a very diverse collection of fishing boats and fishing methods. This is due to the many cultures that have contributed to the North American fleet. Scandinavians, Greeks, Portuguese, Japanese—all have brought the ways of their homelands to the New World. Greek diving methods, Japanese pole-and-line fishing. Italian Monterey clipper boats, Northern European trawling methods, Yugoslavian innovativeness and ingenuity—all have left their mark on North American coastal fishing.

Fishing communities are known for being tight, self-contained little worlds. Fishing is a solitary way of life, and isolates fishermen. Communities whose members share common ethnic backgrounds are especially close-knit. Many such communities in North America still practice the centuries-old customs brought to this continent by the communities' founders. The Portuguese community of San Diego is a good example. It holds an annual festival in honor of Saint Isabel, a patron saint of Portuguese fishermen. Festivals in her honor have been held by Portuguese communities throughout the world since the 1300s.

Immigrants to the New World built communities that resembled those they had left behind. They tended to gravitate toward parts of the country whose climate or terrain reminded them of their homelands. Greeks settled on the shores of the warm Gulf of Mexico, which is similar to the Mediterranean Sea they had come from. Yugoslavs picked parts of Southern California that resembled the Dalmatian Coast of their homeland. Norwegians and Swedes put down roots in more northern locales, particularly around Washington's Puget Sound, an area riddled with long narrow inlets that resemble Scandinavian fjords.

Portuguese

It is said that the Portuguese have an affinity for the sea, that they are at home there, or at least more at home than anywhere else. They have a word, *saudade*, and it is a hard word to translate. It means sadness, homesickness, and a longing to go back to that home. Home is not America. It is not even the Azores, the Portuguese islands that many Portuguese-American fishermen emigrated from. The home, in fact, no longer exists. The *saudade* is an acute longing for perhaps a state of mind rather than a home, and this longing causes pain and sadness. Sailing on the sea is as close as the Portuguese can get to returning to this home. This is as specific as Portuguese fishermen get about the meaning of *saudade*.

The Portuguese seamen who emigrated to the United States in the 1800s did so in an unusual way—by crewing on whaling boats. Whaling skippers of the time were known for their harshness and cruelty. Their reputation became so bad that American seamen would not work for them. Even if you were tough enough to survive the beatings, rigorous working hours and meager rations, you probably would not make much money. Only skippers and boat owners made money. The skipper's motto was said to be: "An aught's an aught, and two is two,/ It's all for me and none for you."

The New Bedford waterfront became a dangerous place to hang out, for when a captain could not raise a crew legally, he was likely to shanghai one. Crimp gangs were paid about $150 to deliver an able-bodied crewman, willing or unwilling, to the captain. Sailors' boarding houses helped in the shanghaiing, and in return were paid blood money. Eventually even the drunks and derelicts became wary of waterfront traps, and the captains had to look elsewhere for crews.

There were places in the world where men lived in such wretched conditions that a berth on a whaling ship was a welcome thing, for it offered a chance for a new life. The Azores and the Cape Verde Islands were such places. They were Portuguese colonies where absentee ownership of land was common, and where tenant farmers were taxed down to the starvation level. Every young Azorian man faced an eight-year term in the military. And the population density was three times that of Europe.

The inhabitants of these islands were whaling captains' dreams. The islanders were used to hard conditions and skimpy rations. They were excellent seamen, they would work for almost nothing, and they were willing to board the whaling ships without even being shanghaied. Starting in the 1820s, whaling captains sailed their boats with skeleton crews from New Bedford to these Portuguese colonies, took on a full crew, and set out for the nearby East Atlantic whaling grounds.

At first, the islanders boarded the whalers only to escape from their native land. Eventually, it became a means of reaching the Promised Land, America. This was especially true of Azorians. Cape Verdans tended to stick with whaling. These tall, tanned, graceful men were superb sailors and quickly rose to minor officers' positions on the ships.

Many of the Azorians went to Provincetown or Gloucester, the two chief Massachusetts fishing ports, after they had served their time on a whaler. Eventually, Portuguese fishermen made up the major part of the Massachusetts cod fleet. Today, the Provincetown and Gloucester fleets are composed largely of the descendants of these Azorian cod fishermen.

When the California gold rush began in 1849, many Portuguese headed west. Men of all nationalities signed onto whaling ships bound for Alaskan waters, planning to jump ship when it passed California. This would give them free passage to the gold fields. A short poem became popular along the New Bedford waterfront concerning this practice: "He who jumps his ship may go to prison,/ But all the gold he gets is his'n."

Talk eventually reached the Azores of the unbelievable riches to be found in the West. Whaling captains, still in need of crews, told the simple fishermen that nuggets the length of a codfish and the girth of a cantaloupe were not uncommon. One leaflet of this time aimed at luring Portuguese trade and settlement to California stated that California doctors charged an ounce of gold a visit and sold their drugs for three or four times their weight in gold.

The whaling captains tried to prevent their crews from jumping ship in California, but many still got away. The Portuguese population in California jumped from 109 to 1,459 between 1850 and 1860. Most of these were Azorians. Most found no gold, and instead established farming communities. These communities drew other Portuguese to California. Most came from New England rather than from the homeland. Some of the newcomers also became farmers, and some became fishermen in the Southern California tuna fleet. By 1930 over 100,000 Portuguese lived in California, almost as many as were in Massachusetts.

Some anglicized their names, so that they were better able to assimilate into American society. For instance, the surname Correia was changed to Curry, Fialho became Fellow, Machado was changed to Marshall, and Dias became Day.

An old Portuguese fisherman who called himself Bacalhao used to live in Provincetown, Massachusetts. *Bacalhao* is the Portuguese word for codfish, and is pronounced "buckle-yow." The man was tall and broad shouldered, and his blue eyes shone brightly out of the gray stubble on his face. He owned a little shanty near the water where he kept his gear. The "fish store," as it was called, was filled with trawling gear, gillnets and other equipment, and had a wide doorway that allowed his dory to be hauled inside and worked on.

One day, old Bacalhao was visited by three fine ladies from Cape Cod high society. Bacalhao greeted them courteously, and wiped the dust off of three backless chairs, so that the fine ladies could sit down.

One of the ladies, who it had been agreed would do the talking, informed the old Portuguese that she and her two friends were *direct descendants* of the crew of the *Mayflower*. Direct descendants on *both sides* of their families. When the lady paused to let the words sink in, Bacalhao murmured politely, "Is very nice." She felt it necessary to make sure Bacalhao knew just what the *Mayflower* was. He confirmed that he did, and she went on.

For the past year, a group they belonged to had been trying to determine just where on Cape Cod the *Mayflower* Pilgrims had come ashore to do their wash. It seemed that the ladies' group had finally found the exact spot where this momentous event took place. Right on the site of Bacalhao's fish store. They had decided that the only fitting thing to do was to erect a monument, a huge boulder, with a bronze plaque on it, in that very spot.

Bacalhao knew all along what the ladies were leading up to. But he decided to play the role of the ignorant foreign fisherman a while longer. "You put a beeg rock, here in thees store?" he asked, looking dubiously at the piles of nets, the corklines, the buoys, the anchors, the tubs of setlines, trying to figure out where such a rock would fit in his little store.

"Uh, I don't think you understand, Mr. Bacalhao," the fine lady said. "You see, your fish store would have to be torn down. We don't expect you to give it up for nothing, of course. We've raised a little fund . . ."

"I hear Tony, down the street, might sell *his* fish store."

"Uh, that won't do, Mr. Bacalhao. The girls have gotten together a fund. I'm sure it will be more than enough to compensate you for your trouble."

The price was far below what his location on the waterfront was worth, as he knew it would be. But he knew there would be no better offer made by these ladies. Their minds were firmly made up.

The discussion continued. The fine ladies drew their fur coats tighter about themselves, for the winter chill could not be kept out of the drafty fish store. Their spokeslady appealed to Bacalhao's patriotic duty. Bacalhao thought back to how he had first come to this country. He had decided to leave his native Azores when he was still a boy. Faced with eight dreary years in the military, he had visited an agent of a New Bedford whaling company, and had signed on. Several days later, he left his home in the middle of the night and stole quietly to the beach, for leaving the country instead of entering the military was illegal. He lit a small signal fire and prayed fervently that he would not be caught. Eventually, he heard the splash of oars. The dory took him aboard the whaler that night, and that was the last he ever saw of his homeland.

He remembered the hard times "furriners" had to go through to make lives for themselves in America. He particularly remembered the struggle he had getting his little fish store.

"Think of the names, Mr. Bacalhao. The first great names in our country's history! Bradford, Brewster, Carver . . ."

Bacalhao had had enough. "You talk of names?" he boomed. He slowly drew himself up to his full six feet four

(Rochelle Smith)

(Rochelle Smith)

(Rochelle Smith)

(Marco)

inches, and loomed menacingly over the fine ladies. "The greatest names in thees country? My name is Bacalhao. You know what that means?"

The ladies did not.

"Codfeesh!" he bellowed. "It means codfeesh! When *Mayflower* came here many years ago, what you think they made for dinner? What you think keep them alive here in New World?"

The ladies didn't know.

"Codfeesh, ladies, codfeesh! Now goodnight!"

A few years later, another committee discovered that the *Mayflower* Pilgrims actually came ashore at a quite different spot than Bacalhao's fish store to do their laundry. Today, a bronze marker, set in a piece of carefully hewn rock, stands on that spot. It is inscribed: "Place of the First Landing of the Pilgrims."

The fine ladies who visited old Bacalhao might have been interested to know of another place where the great name of "codfish" appeared. In 1498, only six years after Columbus "discovered" America and more than a century before the pilgrims landed, Captain John Cabot sailed an English ship to Newfoundland. He found vast numbers of cod in the region, and he found that the Newfoundland Indians' word for the fish was *baccalao*! The Indians must have learned the word from European fishermen, and they must have learned it quite a few years before Cabot's voyage, for it to have become their common word for the fish. The Portuguese have fished off of Newfoundland for hundreds of years, and likely were fishing there long before the Pilgrims set out on their voyage.

Juan Rodriquez Cabrillo, a Portuguese navigator, was the first European to sail a ship into San Diego Bay. Today, a statue of Cabrillo stands on the tip of Point Loma, and watches the tuna boats, crewed predominantly by Portuguese, as they enter the bay and unload their catch at the San Diego canneries. In the early days the tuna were preserved by salting and drying. Large amounts were shipped to the Portuguese and Italian farmers in the San Joaquin Valley, a few hundred miles north of San Diego.

The Holy Ghost, one member of the Holy Trinity, is a patron saint of Portuguese fishermen. Saint Isabel is another patron. Saint Isabel was born in 1271 in Sarragosa, Spain, and eventually became the wife of Dom Dinis, King of Portugal. The land was aflame with civil war at the time. Dom Alfonso, brother of King Dinis, wanted the throne and prepared to take it with force. Isabel, so the story goes, was able to establish peace between the brothers. She was able to bring peace to the country as well, and became known as the Angel of Peace. In 1323 an intrafamily struggle again raged, this time between the king and his son. A fierce battle arose between the two parties at Alvalade. When Isabel heard about the battle, she mounted her horse and rode right into the middle of it. With tears in her eyes, she implored husband and son to cease their hostilities. She succeeded. The next night, the Holy Ghost appeared to her and told her to build a church in his honor as a monument to peace. The day the church was consecrated became a national festival day. For a week before the festival, meat and bread were given to the poor. Portuguese have continued to observe this festival down through the centuries, in many parts of the world. The two focuses of the festival have remained the same: honor to the Holy Ghost and charity to the poor.

In the San Diego fishermen's version of the festival, a fishing boat is selected and its captain becomes president of the Feast of the Holy Ghost. In the festivities that follow, a silver crown and scepter are displayed. The crown stands for both the power of the Holy Ghost and the royalty of Queen Isabel. On the scepter is a dove, symbolic of the Holy Spirit, and of peace. On the Sunday before Pentecost, the crown is taken to the home of the fisherman who has been elected president of the feast, and is left there for a week. The climax of the festival is on Pentecost Sunday. The tuna boats are richly decorated, and a girl is selected by the president to carry the crown from Portuguese Hall to St. Agnes' Church. Bands play and drill teams perform. After a religious service and a banquet, food is distributed to the poor people of the community, and to those who were unable to attend the festival.

In Provincetown, Massachusetts, on the last Sunday in June, a Blessing of the Fleet ceremony is held by the fishermen. This is the same time of year that many fiestas in Europe take place, to honor St. John and to bless fishermen and farmers. For weeks before the festival, fishermen and their wives and children make colorful flags and pennants for their boats.

Festival day begins with a procession from the town hall to the Church of St. Peter. The procession is loud, gaudy and joyful. Both the American and Portuguese flags are displayed. Fishermen carry banners high in the air. The banners are purple and fringed with gold, and each has the name of one of the fishing boats embroidered on it.

In church, they hear a sermon about the beauty and dignity of a fisherman's work. The bishop commends them for taking off during the height of the season to come to church.

In the afternoon, the town assembles at the wharf for the blessing of the fleet. One boat—dedicated to the Virgin Mary—flies a sail with her picture on it, descending from heaven. Other boats fly pictures of St. Peter and other saints.

The bishop blesses the fleet, and the fishermen board their boats, cast off and circle the harbor. One by one they sail past the wharf, where the bishop blesses each one individually.

That night, the bacchanalia begins. It is a joyful, rowdy festival. Fishermen entertain on their boats. Food and wine overflow throughout the harbor. Clams, shrimp, lobsters, wine and delicious Portuguese sweet bread are consumed in great quantities. The merrymaking lasts all night, and in the morning, the boats once again put out to sea.

Blacks

Menhaden boats sailing from Sabine, Texas are captained and crewed totally by Blacks. Many of them have pogy

Fishermen's festival in San Pedro, California. *(Port of Los Angeles)*

fished since they were twelve years old. Most live either in nearby Port Arthur or on the East Coast, and travel to Sabine only for the fishing season. Although whites own the boats, the Black captains are responsible for them and have complete control while at sea.

Black captains say that mixed Black and white crews have been tried in the past, but they haven't worked out. In one Captain's words: "If I hired a white chief, why, we wouldn't get along. It shouldn't be that way, but a white man wouldn't take orders from me. It's even hard mixin' Black and white captains in the same fleet. Hard to get 'em to work together. That's why the Sabine fleet's all Black."

Another Black captain explains why a white man could not take over his job: "Now understand me. Whole lot easier for a Black man to make captain on one o' these boats than a white man. A white man try it, and the crew, which is all Black, say, 'We not gone do shit for him!' So he's laid off in a week, cause he can't produce. If it wasn't for that, why, I'd still be a deckhand."

Although many of the men have been working in pogy fisheries since childhood, only recently were they allowed to be captains. Most of the men were not born in Texas, but came from Florida or further up the East Coast, where they also fished for pogy. It wasn't until 1950 that the Texas menhaden industry began.

Tucker, one of the oldest captains in the fleet, has been fishing for fifty-two years. His father fished too. Tucker lives with his family in Florida during the off-season, but lives on his boat once the season starts. He was one of the first pogy fishermen in Texas, and has been a captain for sixteen years.

Black pogy fishermen follow many of the folk beliefs and superstitions that white fishermen follow, like not turning a hatch cover upside down, or believing that a circle around the moon portends rain. One superstition that

Black fishermen hauling menhaden seine by hand.

seems common only to Blacks is the belief that it is bad luck to kill a sea turtle. Some men say you should not even bring one aboard; others say it is alright to do so, as long as you do not kill it for food. Many Blacks on land hold similar taboos about land turtles. They say that if you kill one, it will come back and haunt you.

In Port Norris, New Jersey, another community of Black fishermen work on the boats that dredge Delaware Bay for oysters. Blacks have been oystering this area since the Civil War.

In recent years, economic conditions in Port Norris have gotten worse and worse. The five hundred Blacks that the boats and "shucking houses" employ are becoming increasingly uncertain about their future. Homes have been abandoned, and more and more people are going on welfare. One reason for the unemployment is that the oyster catch has gone down drastically. The oyster industry lands only one-fifteenth the amount of oyster meat it did a century ago.

Another reason for the bad conditions is that the white owners of the boats and processing plants, in an effort to make the industry more profitable, are replacing men with machines wherever possible. Culling the oysters—separating them from the many pieces of broken shell that also come up in the dredges, and returning the broken shells to the bay—is now carried out by machines. This has greatly reduced crew sizes. Men are still needed on shore to shuck the oysters (separate the meat from the shell), but there is also an effort to replace the shuckers with machines. The traditional method of shucking is to use cracking irons and sharp blades to open the shells and scoop the meat into containers, ready for washing and packaging. In a recent partially successful experiment, the oysters were opened by being tumbled for a while in cement mixers.

Yugoslavs

There is evidence that in the sixteenth century a sailing vessel from Croatia, in what is now Yugoslavia, sunk off of the present-day North Carolina, near the Roanoke colony founded by Sir Walter Raleigh. Survivors of the wreck made their way to a nearby island inhabited by a tribe of Indians. The sailors stayed for the rest of their lives with this tribe. They married Indian women and assimilated into the tribe. Eventually, these Indians became known as the Croatans. Croatan Indians still live in North Carolina, as well as in neighboring states.

In the four hundred years since the shipwreck took place, most of the details of this mingling of two cultures have been lost. A check of old Croatian sailing records, though, reveals that a ship named the *Croatan* did set sail at about the time that this event was supposed to have taken place.

Large-scale Yugoslavian immigration to North America did not begin until late in the nineteenth century. Many Croatians and Slovenians settled in Michigan and Minnesota, where they worked in the copper and iron mines. Slavs from the Dalmatian coast, on the other hand, settled near the ocean, for they had long traditions of earning their livelihood from the sea. For centuries, Dalmatians have been renowned as excellent mariners and shipbuilders. Many Dalmatian immigrants settled in the Mississippi Delta region, where they started a lucrative oyster industry, and on the Gulf of Mexico, where they worked in the rich gulf fisheries. The oyster industry they started is much like the one found today on the Peljesac peninsula of Yugoslavia.

The 1849 gold rush attracted large numbers of Dalmatians to California. Many of them who did not get rich mining moved to the Los Angeles area, where the climate and topography resembled their native Dalmatian coast. They opened grocery stores, restaurants and saloons, and entered the San Pedro fishing fleet, eventually becoming the dominant ethnic group in that fleet. Dalmatians settled and fished on all parts of the West Coast, from Southern California to Alaska. Quite a few left the Columbia River and Alaska salmon grounds and migrated south to fish with their countrymen in the growing San Pedro fleet.

Today, San Pedro is one of the world's foremost fishing ports, producing more fish than the Massachusetts ports of Boston, Gloucester and New Bedford combined. Each year a large Fisherman's Day festival is held. Before the Pacific sardine fishery became defunct, the day for the festival was picked to coincide with the break between tuna and sardine season late in September, when the fishermen were all in port changing their gear. Sardine are hunted at night, and are spotted by the phosphorescence they give off. It is easiest to spot them during the dark nights of a new moon. When the moon is full, it is generally too bright to locate schools of sardines. Thus, the festival was, and still is, held on the full moon.

The festival culminates with the Blessing of the Fleet, given in Slavonian as well as in English. A wreath of flowers is tossed into the sea in memory of the dead. Fishermen compete in rope splicing and net mending contests.

One of the important events of the festival is the high mass held in the old Mary Star of the Sea church. The church is lavishly decorated with gladioli and chrysanthemums. Over the altar are written the words *Maria Stella Maris Ora Pro Nobis* ("Mary Star of the Sea, Pray for Us"). After the mass, the cardinal leads a procession to Fisherman's Dock for the Blessing of the Fleet. The boats have been bountifully decorated for this occasion. Four fishermen carry a statue entitled "The Holy Image of Mary Star of the Sea." The cardinal blesses the fleet, after which roses are thrown into the waters in memory of fishermen buried while at sea, and the prayer *De profundis* ("Out of the Deep") is recited. Then the long parade of boats begins, amid much blowing of sirens. Thousands of dollars of prize money are given away for the best-decorated boats. The decorating is often done in secret, sometimes in hidden coves of Santa Catalina Island, twenty-four miles southeast of the harbor.

Some of the most important innovations in the North American fishing fleet were introduced by Yugoslavs. Shortly after World War I, Peter Dragnich and Martin J. Boganovich introduced the first boats on the West Coast capable of storing fish for long periods of time through the use of ice. Years later John Zuanich experimented with preserving fish in brine, a method used today by most large tuna boats. In recent years Nick Bez of Seattle experimented with processing seafood on large floating canneries that accompany the fishing boats. The invention that has had the greatest effect on fishing methods in the last two decades is the power block, invented by Mario Puretic while he was working in the Southern California sardine, mackerel and anchovy fisheries.

Italians

The old man was patiently mending a drag net when I first talked to him. His hands were gnarled like twisted oak roots, and his face was weatherbeaten, and looked a bit tired with all the life it had seen. He told me he had only been mending nets for a week—he had just been hired. His hands worked too skillfully and effortlessly for me to believe that. I asked a younger man who was working near him about it. He laughed—a strong, loud laugh. "He told you that?" he said. "Why, that man's the original Godfather! He worked on every type of fishing boat you can name for seventy years. He taught me all I know. I'm his son."

I went back to the old man. He did not want his picture taken, for he felt a bit of his soul would be trapped in the camera. I asked him to show me how to mend the net. "What you do for me?" he barked. I told him I was a teacher. He let out a gasping wheeze, which was his version of a laugh. "What can you teach me? English, maybe? You teach me English, I teach you how to mend nets." He eventually became friendlier, and talked about his life. He came to the United States when he was a small boy, with his father, who had owned a fishing boat in Italy. They bought a boat in this country and gillnetted together near Atlantic City, New Jersey. Later, they fished on Georges Bank, off of Massachusetts.

One time, his boat hit a buoy and started to sink. In those days, boats had no radios with which to call for help, so he had to stay in his boat, which was half sunk and half floating, for two days until another vessel happened by.

Later, he moved west and fished Alaskan and Californian waters. Today, he owns three drag boats. He does not go out anymore. He says he is "too old to take it."

He doesn't like America. "Everybody tries to kill each other here. Back in Italy, nobody had much, but they helped each other. And somehow, everyone had enough. Here, everyone's out for themselves. They'll cut your throat." Who "they" are, he would not say.

There is a family of fishermen in Gloucester, Massachusetts whose history is similar to that of many Italian fishermen's families on both the East and West Coasts. The family's patriarch came to the United States in the early 1900s and worked for a while on the transcontinental railroad. After this, he made his way to Gloucester and got a temporary job on a gillnetter. One day, the boat's rudder broke. He jury-rigged one from a pine box. This got him a permanent position on the boat. He saved enough money to buy a forty-two foot boat of his own. He worked it efficiently enough to buy larger boats, which he and his six sons used for purse-seining mackerel. He taught his sons to fish during their vacations from school. When they graduated from high school, they worked with him full time. His sons eventually acquired their own boats and went their separate ways. Two of his sons and two of his grandsons opened a commercial fishing gear store. Today, the family still owns the store, as well as five fishing boats.

Many of the Italians who worked on the construction of the transcontinental railroads migrated to California when they were through with their jobs. In California, they worked in the vineyards and orchards, and as fishermen. The San Francisco fleet is today predominantly Italian, and there are many Italians in the San Pedro, California fleet too.

San Francisco's Italian fishermen hold a festival each year on the first Sunday of October to pay homage to their protectress, *Maria Santissima del Lume* ("Our Lady Mary of Light"). The festival is smaller than some other festivals, for it is pretty much for fishermen only, rather than for tourists as well.

Maria Santissima del Lume is also protectress of Porticello, a village near Palermo, Sicily. There is a legend that many years ago, a Porticello fishermen found a statue of incredible beauty washed up on the beach. On the statue was written: *Maria Santissima del Lume*. He placed the statue on the altar of the church. It disappeared the next day, and was found back on the beach. Twice more, it was transported to the church, and twice more it appeared back on the beach. It was decided to leave it there and build a church around it. The Sicilian fishermen of San Francisco have a large painting of Maria Santissima del Lume, and keep it in the Church of SS. Peter and Paul. After this, there is a procession down to Fishermen's Wharf. No queen is elected during the festival, for Our Lady Mary of Light is *the* queen. At the wharf, the pastor of the church blesses both the boats and the sea. A bazaar is held for the rest of the day, and a festival atmosphere permeates North Beach and Fishermen's Wharf.

During the last days of August, the Italian fishermen of San Pedro, California hold a triduum—three days of prayer. They pray to their patron, *San Giovan Giuseppe della Croce* ("John Joseph of the Cross"), an eighteenth-century priest from the island of Ischia, off of Naples. Many of the old fishermen of San Pedro are from this island. Their boats often contain a shrine and statue of Saint John Joseph.

Italian immigrants who came to the United States in the 1800s brought with them the fishing methods they were accustomed to in the Old Country. Their method of trawling was to use two boats for every one net. Each side of the large, funnel-shaped trawl net was attached by a cable to one of the boats. Instead of otter boards or a beam, it was the distance between boats that kept the trawl mouth open. The two boats had to travel at exactly the same speed, in parallel lines to each other, if this method was to work. The two captains and crews had to coordinate the movements of their boats very closely, and split the catch at the end of the day. Perhaps it was found that one vessel could catch almost as much as two vessels if it hauled the net alone, for out of this method evolved a system where the two cables from the net went to two different spots on one vessel, instead of to two separate vessels.

Greeks

Six days out of every year, the Greek sponge fishermen of Tarpon Springs, Florida, do not dare to leave port. These are the days between New Years and Epiphany Day, the sixth of January. The sea is believed dangerous to sail upon during this time, for the waters are not blessed.

The sea does not become safe until a priest throws a golden cross into it on Epiphany Day. The moment the cross hits the water, young divers leap in after it, in an effort to be the one who finds and retrieves it. The one who retrieves it receives a special blessing, and is supposed to enjoy good luck for the rest of his life. After the cross has been found, the priest blesses the whole sponge fleet. This ceremony climaxes three days of Greek dancing, singing and feasting. Days before the blessing of the fleet, the young divers visit homes in the Tarpon Springs community, seeking donations for the church. The members of the Greek community give generously to the divers for it is considered good luck for the winning diver to cross their thresholds.

In addition to blessing the waters, priests bless each boat every time it is ready to leave on a trip. The priest dips a cross into a pan of holy water and sprinkles it on the deck, the masts, in the crew's quarters and even on the engine. He also blesses the captain and each of his crew. It

Sponge boats. *(Tarpon Springs Chamber of Commerce)*

is believed that this ritual not only protects the boat from harm, but also insures a large catch.

Priests are respected and important members of the community. Under certain circumstances, they are also feared, especially when run into unexpectedly. Women often return home and set out later in the day if they see a priest on the streets early in the morning. Children tie a knot in their handkerchief if they meet a priest on the way to school. Some men touch their genitals to protect their virility. Captains will postpone their trip if a priest appears unexpectedly on the docks. Elias, an old Tarpon Springs skipper, tells the story of how a priest came back from a fishing trip just as Elias was about to take some friends on a pleasure cruise. He wanted to cancel the trip, but his friends wouldn't let him. Some of the party went swimming, and sure enough, they got caught in a rip tide and almost drowned. Greeks of Tarpon Springs say that misfortune is inevitable if a priest is seen right before the beginning of a trip.

Captains will postpone a trip if a funeral has just taken place, too. "When they break ground for a dead man, it is bad luck. On that day, nobody leaves." If a captain is about to sail and somebody asks him when he is going to leave, he will delay his trip. Women in mourning clothes are unlucky if spotted as a boat is leaving. One old sponger tells the story of the woman who always wore black, even before her husband died. She repeatedly showed up at the docks, as if she wanted to lay a curse on the men there. And each time she came, something would happen. A diver would drown or a boat would break down. Finally, the priest had to tell her to stay away. It seems that those who have "touched death," like priests or widows, have gained the power to bring misfortune upon other people.

Sponge fishermen wear golden crosses or pin charms to their clothing to ward off evil. Sometimes they carry pieces of consecrated bread or containers of holy water.

The boats carry icons of Saint Nicholas, the patron saint of Greek fishermen. It is said that this saint has the power to control the wind and subdue a storm. If the sea is dangerously rough, the icon is dipped in the water to calm it. At other times, oil from the lamp that burns in front of the icon is poured on the waters.

Many captains carry icons of their families' patron saints as well as the Saint Nicholas icon. One fisherman tells the story of how his father lost his rudder during a storm, and got on his knees and prayed in front of an icon of Saint Catherine, his patron saint. Soon after, he heard a beating on the side of the boat. It was the rudder, bumping against the outside of the hull. He reached down and hauled it aboard, and was able to make it home safely. For the rest of his life, he sent the money he earned on the last day of every trip to Saint Catherine's monastery on his native island of Calymnos. If he had found a solid gold sponge on the last day of a trip, he would have sent it to the monastery.

If a Greek sponger sees a waterspout while at sea, he will carve a cross on the mast and throw a knife into its center to insure that the spout stays away from the boat. This practice is in violation of the laws of the Greek Orthodox church, though, and the man must do penance afterwards.

Many fishermen from other parts of the country feel that it is bad luck to leave on a trip on Friday, the day that Christ was crucified. Greeks from Tarpon Springs have a similar taboo, but it concerns Tuesday, not Friday. Tuesday was the day that Constantinople fell to the Turks in 1453, and it has since been considered a day of bad luck. (Other groups, such as southern Blacks, also have a Tuesday taboo, which suggests that there may be some explanation other than the fall of Constantinople.) Women are told that if they start making a dress on Tuesday, they will never finish it. Journeys of any sort, including fishing trips, should not begin on Tuesday.

Greeks have lived in Tarpon Springs for seventy years. The community was started by John Cocoris, a sponge wholesaler. He visited the town around the turn of the century, and noticed that the whole sponge industry depended only on what could be caught close to shore by the hooker boats. Hooker boats snag the sponges with three-pronged steel hooks attached to long poles. They work beds in waters from two to fifteen feet deep. The rich deepwater sponge beds were untouched, for no one in town had the equipment or the know-how to harvest them. Cocoris imported the necessary diving equipment and a crew of professional divers from Aegina, an island near Athens. The crew was so successful that two years later in 1907 there were 800 Greek fishermen in Tarpon Springs, operating fifty diving boats and fifty-five hooker boats.

Sponges are marine animals. After they have been carefully brought up to the boats by either divers or hooks, they are laid on the deck for two or three days to die. Then they are returned to the water and the black skin that covers them is beaten off with pieces of wood. After this, they are dried to prevent spoilage. They are cut into uniform sizes and graded according to firmness and freedom from tears. Natural sea sponges are more absorbent and easier to clean than synthetic ones.

The booming village of Tarpon Springs became known up and down the coast as "Greek Town." Most of the immigrants came from the Dodecanese Islands, of which Aegina is one island. Their families had always earned their living from the sea, and this tradition was continued in the New World. Greek fishermen settled in other parts of North America like Provincetown, Massachusetts or Deas Island, British Columbia. In both of these places, they continued to work at their trade. Most of the Deas Island Greeks came from Scopelos in the Aegean Sea. Immigrants from the Greek mainland followed different paths in the New World. They gravitated toward the large cities, and few became fishermen.

Tarpon Springs has retained more community solidarity and more of the Old World traditions and superstitions than most other Greek communities in North America. Greek captains in Tarpon Springs employ Greek crew members and divers on their boats. Greek businessmen and storekeepers depend heavily upon other Greeks in the community for their livelihood. If a diver dies at sea, the whole community mourns his loss and consoles his family. If a sponger's boat sinks, the community helps him pay off his debts and get back into business.

The town's immigrants still use Greek as their everyday language. Most of their American-born sons and daughters are bilingual and speak Greek as well as English. Most of the businesses in town are owned by Greeks. And there are eight ethnic organizations in town, for Greeks of all ages and both sexes, that are dedicated to the same goal—to keep Greek culture alive.

Cubans

While the sponge industry on the west coast of Florida is controlled by Greeks, on Florida's east coast it is dominated by Cubans. Most of the spongers themselves are Cuban, and the Cuban buyers have a virtual monopoly in the area.

It is surprising that there had never been a sponge marketing center on the Atlantic coast of Florida sooner than 1950, when Cubans started entering the field, for sponging has been an important part of the area's economy since 1850. Until the 1900s the catch was brought to Key West. Early in this century, Tarpon Springs, on the west coast of Florida, surpassed Key West as the world's sponging center, giving east coast spongers a choice of markets.

In 1915 some Miami businessmen became interested in opening a sponge market in Miami to rival the one in Tarpon Springs. Perhaps their plans were forgotten when the Florida real estate boom began several years later. Whatever the reason, the sponge market did not get started until 1950, when a large sponge dealer from Batabano, Cuba, Andres Dworin, emigrated to Florida and started the East Coast Sponge Company. Shortly thereafter two Cuban brothers, the Arrelanos, also from Batabano, started a sponge packing house on the Miami River.

When Castro came to power in Cuba, many Cubans emigrated to Florida. Among them were men who had sponged all their lives, and knew no other type of work. By 1960 some of these expatriated Cubans had entered the sponge fisheries of the Florida Keys and Biscayne Bay. By 1968 fifty Cubans were engaged in one aspect or another of the east coast sponge industry.

When the two Arrelano brothers first set up their packing house, they made weekly trips along Florida's east coast in their two-ton truck, buying sponges from Cuban fishermen. They also bought from Americans who used to make the long trip to the Tarpon Springs markets. Today there are few spongers along the east coast who do not sell to the Cuban brothers' company. The company, in turn, markets the sponges throughout the United States.

The east coast market has grown, but it is still much smaller than the one in Tarpon Springs. This is largely because the Rock Island wool sponge found near Tarpon Springs has for years had the reputation of being the softest and most absorbent sponge in the world, even though government-sponsored tests have shown that east coast sponges are equal or superior in quality. Sponge wholesalers have traditionally bought their stock from Tarpon Springs, and traditions change slowly in the sponge industry.

Many Cuban-Americans work in the lobster fisheries off of the Bahamas. The government of the Bahamas recently declared that the lobsters off its shores were creatures of the Bahama continental shelf, and forbade non-Bahamans from harvesting them. This action put many Cuban-Americans, among others, out of work. Many of the Cubans have since lost their boats, cars and homes, for other jobs have been hard for them to find. Some of them are in their fifties and sixties, and know no other type of work.

Scandinavians

When the railroads came to the Pacific Northwest in the 1870s and 80s, they did extensive advertising in other parts of the country on the beautiful land that could be homesteaded in the area. They hoped to draw hordes of settlers to the West Coast, and make large amounts of money from train fares. Their advertising campaign was quite successful. In 1883 one hundred immigrants per day were arriving in Seattle. Most were Norwegians and Swedes. At first they worked as loggers and fishermen, and eventually turned to dairy farming and poultry raising. Some men did all of these things at different times of the year. Many settled around Puget Sound. The area probably reminded them of their native Norway, for Puget Sound has many long and narrow inlets that resemble Norwegian fjords. The steep, snowy mountains around the sound also resembled their homeland. And the work they found—fishing, logging and farming—was work they had done at home.

Although the greatest density of Scandinavians on the West Coast is in the Puget Sound area, many settlers made their homes on other parts of the coast as well, from Los Angeles up to Alaska. Many worked in the rich salmon fisheries of the Columbia River. Although they had fished extensively in their home countries, few had ever fished salmon before, and so the methods and skills of this fishery had to be learned from scratch. In 1875 many of the men fishing the Columbia were employed by the salting and canning plants. They worked two to a boat (one was the rower and one the fisherman), and got paid four cents for every fish they gillnetted. If a man was lucky enough to own his own boat, he could make more. In those days, a good-sized salmon could be bought retail for fifty cents. Men from all the Scandinavian countries came to the Columbia. In the 1880s, the most numerous group were the Finns. Swedes also were plentiful in this area. There were many Finns to be found south of the Columbia, too, in places such as Coos Bay, Oregon.

Scandinavian fishermen of Puget Sound, circa 1900.

Catching the "holy flounder," halibut, has been the task of Norwegian-American fishermen for a century. In the early years of this fishery, 95 percent of the North American halibut fleet and 99 percent of its boats' skipper-owners were Norwegian. The fleet is still largely Norwegian, although its size has shrunk in recent years, for halibuting is very rigorous work (perhaps the most rigorous type of fishing) and catches are not always that good.

Norwegians have not limited themselves to halibuting. The Pacific Northwest's herring and albacore fishermen are largely Norwegian, and most of Washington's Puget Sound salmon trollers are run by men of Norwegian ancestry. The sound's salmon purse-seining operations, though, are dominated by Slavs.

Each year, large numbers of Washington's salmon boats trek northward to the rich salmon grounds of Alaska. This trek is reminiscent of the yearly journey of cod fishermen in Norway from the southern part of the country to the cod spawning grounds around the Lofoten Islands, in the north.

Poulsbo, Washington was settled by Norwegians in the latter part of the nineteenth century. It is still a Norwegian fishing village whose men are active in the salmon and halibut fisheries. Every year, Poulsbo celebrates *Syttende Mai* ("the seventeenth of May") with dancing, songs and native costumes. The seventeenth of May is the Norwegian day of independence. It was the day the constitution was accepted declaring that Norway was free of Swedish rule.

In the 1880s a Norwegian fisherman named Hans Helgeson, of Victoria, British Columbia, rigged up a sloop and sailed north to the Queen Charlotte Islands, in search of new fisheries. He heard stories from the Indians about a delicious type of fish that was completely unknown to the white man. Helgeson got an Indian chief to take him in a canoe to the place where the fish were found. They sunk lines down to three hundred fathoms, and caught fish similar to, but larger than, the cod he caught in his native Norway. He dubbed the fish black cod, and found that they existed in great quantities in the area. A few months later, he fitted out a schooner with a Norwegian crew and returned to the area, loaded up on black cod, and sold them in Victoria. In a short while, other boats entered the fishery. In 1889 a colony of Norwegians was started at Sanders Harbor, British Columbia, to fish for and process black cod. They marketed the catch mainly in Australia, for they found that North Americans were not very eager to eat the fish. Although black cod has continued to be caught until the present day, it has never been that popular as a food fish. In recent years, with more traditional food fish catches like salmon and cod steadily decreasing, the black cod fishery has started to grow.

Many Norwegians settled even further north and opened salmon, cod and halibut fisheries in Alaska. They noticed, and wrote about, the many ways Alaska was similar to Norway. The same type of berries grew there, fishing was an important part of the economy, summer days were endless, and the climate in Southeast Alaska was very much like Norway's, for Southeast Alaska is tempered by the warm Japanese Current in much the same way as Norway is warmed by the Gulf Stream.

Scandinavians have not limited their fishing activity to the West Coast. Much of the fish that is sold at New York's huge Fulton Fish Market is caught by Norwegians. They played a significant role in the East Coast whaling industry of the last century. Many are also involved in oyster harvesting in Chesapeake Bay and in the fisheries of the Gulf of Mexico. They are found all over North America, but most Scandinavian fishermen work out of ports in the Pacific Northwest.

Native Americans

Archeological finds have indicated that Indians in the Pacific Northwest have been fishing salmon for 9,000 or more years. Salmon used to be the mainstay of Pacific Coast Indians' diets. They caught a year's supply during the salmon's spawning runs, and dried and smoked it for winter use. The ample food supply allowed the coastal Indians time to develop their cultures, and become very fine artists. Dried salmon was used as currency as well as for food.

The catching of the first salmon each year was an event celebrated with feasting and dancing. A salmon's soul was considered immortal. It was believed the fish voluntarily let itself be captured, in order to feed its friends, the Indians. The Indians, in return, revered the salmon, and gave it names like "Great Jumper" and "Lightning of the River." They had strong taboos against molesting the fish except during the traditional fishing times. It was extremely bad luck to poke its eyes out. This could bring tragedy upon the tribe, for the salmon might refuse to return to the river to spawn.

The Indians had many ways to catch the salmon. The Tsimshians of Canada's west coast built stone walls that were just low enough to allow the salmon to swim over them with the incoming tide. Then, as the tide ebbed, the fish would be trapped in the shallow pool just upstream of the stone dyke, and could be easily speared, or picked up by hand, if the water was low enough. The salmon were brought to the women, waiting on shore, who cleaned and sliced the fish into thin fillets. These were laid on racks to dry, and then put into the smokehouses. The smoked and dried salmon could be kept for two years.

Other tribes caught the salmon in cedar traps or in dipnets, or, if the run was particularly heavy, went out in their canoes and beat them over the head with canoe paddles. Some tribes built rock jetties or rickety wooden platforms that protruded into the rivers below waterfalls or rapids, and speared or dip-netted the fish as they attempted to climb the rapids. Others made barges by lashing planks between canoes, and harpooned or netted the fish from them.

For many years, the native peoples of the North Pacific coast—Niskas, Gitksans, Carriers, Chilcotins, Shuswaps, Kwakiutl, Tsimshians and others—had enough salmon for food and for barter. Then came the white man. The Indians soon found themselves fishing for trinkets and whisky instead of for food. Soon, there were noisy, smoky canneries dotting the shores of many rivers, and more and more of the Indians became low-paid employees of the white man, instead of independent fishermen. They saw their rivers ruined by pollution, overfishing, and erosion from hydraulic mining and logging. They saw countless spawning runs blocked by dams. Catches steadily declined. There are still old Indians who remember the days of plenty, but these days are drifting further and further into the past.

Native Americans certainly have much to be angry about. The issue, however, is not completely one-sided. There are certain areas where whites, too, have legitimate gripes. It has been said that every treaty made between the United States and the Indians has been broken, usually by the white man. This was not the case February 12, 1974. On that day, Federal Judge George Boldt ruled that the spirit of certain Indian treaties signed in the years 1852 to 1856 had to be obeyed. What he interpreted this to mean was that the 15,000 of Washington's Indians affected by the treaties are entitled to half of all Washington's harvestable salmon. And they have the right to catch this salmon using any type gear they choose. They can even use fish traps, which have been outlawed in most other salmon fisheries. Boldt's decision gave 50 percent of the state's rich salmon resources to only 2/10 of 1 percent of the population.

The decision has caused an uproar not only among commercial fishermen, who fear for their livelihood, but among sport fishermen as well. The steelhead, a type of rainbow trout and the most popular Pacific Northwest gamefish, is a member of the salmon family and thus comes under the 50 percent ruling. The sportsmen are doubly annoyed with Boldt's decision because hatcheries for this fish are funded mostly out of sportsmen's fishing license fees.

The only rule the Indians must follow is to keep detailed records of their catch. They are not limited to fishing on their reservations, but can fish in the traditional off-reservation areas that their ancestors used for centuries. No one has figured out how state authorities can determine when the Indians have landed their 50 percent of the fish. In 1973, only about 700 Indians, most of them gillnetters, fished Washington's off-reservation waters.

Canadian fishermen are worried about the decision, too. Much of the sockeye and pink salmon caught in Puget Sound come from the Fraser River in Canada. The catch is carefully divided by the American-Canadian International Pacific Salmon Fisheries Commission. Unrestricted Indian fishing could upset this division. In retaliation, Canadians might completely fish out the salmon runs before the salmon have a chance to reach United States waters.

Commercial and sports fishermen have often warred with each other in the past, but now have entered into an alliance to combat Boldt's decision. Some of the their demands are to protect the steelhead by making it a national game fish, and to limit the Indians to a percentage of salmon more in keeping with their percentage of Washington's population.

By May 1975 the joint Canadian-American fishery had not yet been affected by Boldt's decision. Nor had coastal fishing areas. But the white man's Puget Sound net fishery was all but destroyed when the 1975 season for whites was shortened to only four days long.

Boldt's decision was appealed, but on June 4, 1975, the Eighth Circuit Court of Appeals upheld the decision. Washington's Attorney General announced that he would bring the appeal to the United States Supreme Court.

Most of Washington's white fishermen make their best catches in the northern waters of the state, the waters that are fished jointly by the United States and Canada. In September 1975 Judge Boldt ruled that the Indians should have a chance to take 50 percent of the American catch in those waters, too. He allowed them to fish mixed gear, which gives them an advantage over whites. Whites can only use purse seine gear during the day and gillnets at night. Indians can use either, night or day.

White fishermen have staged several demonstrations against these rulings. In one, several hundred gillnetters staged a huge "fish-in" in Puget Sound during a time when they were not allowed to fish. Ninety-eight violations were handed out, but charges were quietly dropped later. In another demonstration, 300 fishermen surrounded the federal courthouse in Tacoma and displayed signs such as "Boldt is an Indian Giver."

The Quillayute Indians of Washington fish smelt, a tasty fish similar to salmon but much smaller, from dugout cedar canoes in much the same way their ancestors did a millenium or more ago. Many of the canoes now have outboard motors mounted on them, but the motors are usually not used when taking smelt, for the paddle has proved to be a more effective means of locomotion for this type of fishing.

They fish with beach seines about twenty fathoms (120 feet) long and one fathom deep. Their fishery is in the Quillayute River, formed by the junction of the Dicker River with the Soleduck, Calawah and Bogachiel Rivers. The waters are fed by glaciers in the Olympic Mountains.

Different families have their traditional fishing spots, located on spits jutting into the river. When the smelt are about to begin their spawning run, the families go to their various spots, light fires, and sit and wait. When the smelt first appear somewhere downriver, the Quillayute families do not jump up and run to the spot, for this would encroach upon the ancestral grounds of some other family. They patiently wait for the smelt to come to them.

When the smelt finally arrive, the Quillayutes still do not rush. They have fished smelt for centuries, and know that too much splashing and activity can spook the little fish. They get into their slim cedar canoes and slowly and smoothly set their nets, which have been neatly piled in the canoe's stern. If an outboard motor is used, it is hung on a bracket on the side of the boat, or placed in a well just forward of the stern.

They set their nets so that the smelt will swim into them as they head toward their spawning grounds. The engine is usually not used, for the net can easily foul the propeller. Their paddles provide ample power over the short distances the canoes must travel. After the smelt are just about encircled by the seine, the canoeman uses his intuition and experience to decide how long to wait before closing up the net and hauling it in. More fish can enter if he waits. Fish will escape, though, if he waits too long. When the canoeman decides it is time, he brings the towlines to the Indians waiting on shore, who haul in the seine. A good set will bring in 300 to 400 pounds of fish.

Japanese

Japanese-American divers used to dominate California's abalone fishery, prying millions of pounds of the mollusks off submerged rocks each year. Their boats used to comprise an important part of the large Southern California tuna fleet, too. But all this was before World War II. In the months after Pearl Harbor, Japanese-Americans were sent to concentration camps to sit out the remainder of the war. Their boats and gear were confiscated, and their jobs in the tuna canneries were taken by Slavs and Italians. The tragedy is that after they were finally released from the camps, they could not go back to fishing, for their boats were owned by others. The abalone fishery had been taken over by Caucasians, many of whom had learned to dive during the war. The last thing they wanted was to let the Japanese work with them, for anti-Japanese feelings still ran high.

The Japanese entered the California fisheries in 1901. A group of Issei railroad workers, in San Pedro on holiday, noticed the large numbers of *awabi* (abalone) that could be found under rocks in the intertidal zone and in deeper waters. *Awabi* is a delicacy in Japan, but at that time was barely know to America. The railroad workers decided to set up a dried abalone business in San Pedro. The residents of the area were very hostile toward them, though, and so they moved to nearby Terminal Island, building their houses on piles driven into the tideland mud. They branched out into other types of fishing, and by World War II, owned and operated a fleet of tuna boats that roamed all over the Pacific. Five hundred Japanese families lived on Terminal Island at that time.

Immediately after the Pearl Harbor attack, the Terminal Island fleet was recalled, for it was felt that they were a threat to United States security. Within hours, most of the smaller craft arrived. The larger ones—some a thousand or more miles away—could not get back to port for several days. Many fishermen were badly beaten as they stepped off the boats by hysterical mobs who had been waiting at the docks for them. The canneries fired all Orientals, and hired Slavs, Italians and Filipinos in their place. The many Japanese farms and floral gardens were taken over by Caucasians.

When the Japanese were finally released from the concentration camps after the war ended, they found that their homes on Terminal Island had all been destroyed. Few people lived on the island anymore, and those who did were either Army personnel or convicts. They were able to regain some of their farms, but were often driven off by irate neighbors. Although the Japanese community has, through years of struggle, become fairly well-off financially, few have been able to reenter the West Coast's fisheries. All that remains of their once-thriving tuna fleet and abalone industry are memories.

Appendix

FISHERMEN'S SUPERSTITIONS

If a Greek fisherman sees a waterspout, he often carves a cross on the mast and hurls a knife into the center of it to keep the spout away. Black fishermen throw quarters or half dollars over their shoulders to "buy up" a certain amount of wind. This custom originated in the days of sail, but is still used today when crews are overworked. A man will toss some silver coins into the sea in the hope that a storm will arise, giving the tired crewmen some time off.

Silver dollars are sometimes put underneath the mast when a boat is built to give it good luck. Many men wear gold charms that have been blessed by a priest to bring them luck. One Greek sponge diver told of how he refused to dive during one of his trips because he had forgotten his lucky charm. Later on in the trip, he substituted a piece of string tied in the shape of a cross for his charm, and resumed his normal duties.

The appearance of a porpoise is good luck, and there is a taboo against killing them. Black fishermen have a similar taboo against killing a sea turtle. It is bad luck to see an albatross or hear a loon cry. Saying the words "alligator" or "pig" also invites bad luck. One should never whistle on board a boat, for it is likely to bring a gale. And if you open the hatch, never turn its cover upside down.

Some boats are jinx boats, no matter what is done to fix them. One vessel called the *Carolina Cape* was such a boat. The owner had to put two new engines into it in less than two years. He finally sold it, and the new owner had to put a third engine into it the first year he got it. Fishermen avoid crewing on hard-luck boats, for they know that there is no money to be made on them.

A lamb stew called *kavourmas* is often taken along on Greek fishing boats. It is packed in sealed jars, and there is a strong taboo against opening these jars before the boat leaves port. One man who did so reportedly saw the face of death in the stew, and died on that trip.

Many Black fishermen tell a tale about a white captain who would curse God every time the fishing was poor. Once he took a hatchet and climbed the mast, daring Jesus Christ to meet him halfway and give him a fight. His crewmen thought it was because he was wicked that he didn't catch anything.

Many fishermen do not believe in the above superstitions. Others say they don't believe them, but obey them anyway, just to make sure. One Texas fisherman explained it like this: "You have so much bad luck in this business, there's no use temptin' nothin'!" Superstitions are more widespread among those who fish on the open sea than among those who fish in safe, sheltered bays. On the open ocean, the weather is unpredictable and potentially more dangerous and the number of fish that will be caught depends more on chance than in quiet, protected bays.

HARBOR HIERARCHIES

I've fished out of this harbor for twenty-eight years, and I know every skipper and boat in it. Harbors are little cities. They have their high-class residents and their bums. See that group of seventy- and eighty-foot draggers and crabbers over on the first float? They're the upper class of this harbor. They're all steel-hulled boats. Some cost over $300,000. Today, they would be even more expensive. Their skippers can afford the best equipment made. That's why most of the harbor's highliners are from that fleet.

Next comes our "upper middle class." The boats are a bit smaller. Some are steel hulled, and some are wood. Their owners drive fancy cars, live in nice houses, and are deep in debt.

I'm in our middle class, I guess. I own a thirty-five foot troller and I run it singlehanded. I like being the only one on the boat. That way I don't have to please anyone but myself. If the fish aren't biting and I feel like coming in, I don't have to argue with any deckhand. The bigger boats generally don't go out singlehanded. Some have one crewman aboard besides the captain. Some have more. That big crabber tied up by the harbor master's office takes five men on each of its trips.

We've got some fine boats in this harbor, but we've got some dregs, too. I could show you some boats that are half eaten through with dry rot, or so cluttered with garbage that it's disgusting to walk by them. They're like that because their skippers don't care. A boat reflects a man's values. A guy will buy a boat, and in six months, it will look completely different. He might put a new paint job on it, change its gear, do a whole bunch of things to it, and soon it becomes an extension of his personality. Jim Parker just bought that drag boat at the end of the float, and the first thing he did was paint a big picture of a naked mermaid on the deckhouse. That tells you something about Jim.

There's one more fleet I haven't told you about. See that old float across the river, the one that looks like it's going to sink at any moment? It's been patched up so many times, it's a wonder it's still above water. That's where the hippy fleet ties up. They've got every type of boat you can name, from twenty-foot dories to $65,000 trollers. They seem a bit more relaxed about their work than most of the guys around here. More "mellow," they call it. Some are so relaxed, they never seem to make it out of the harbor. Others are pretty serious, and are fine, hardworking fishermen.

Glossary

AFT. Toward the stern.

AMIDSHIPS. In or near the middle of the boat.

ANADROMOUS FISH. Fish like salmon or steelhead trout that live in the sea but ascend rivers to spawn.

BALLAST. Weight carried inside the boat or outside on the keel for trim and stability.

BEAM. The boat's width at its widest point.

BEAM TRAWL. A conical-shaped net held open by a horizontal beam. At each end of the beam are iron frameworks that hold the net open in a vertical direction.

BLOCK. A wood or metal shell that encloses one or more pulleys.

BOTTOMFISH. Groundfish.

BOW. The forward part of a boat.

BRAIL NET. A small dip net for scooping out portions of the catch from the main net and hauling them aboard. They are used to transfer tuna, salmon and sometimes menhaden from the purse seine to the boat's hold.

BULKHEAD. Any wall in a boat.

CAULKING. The compound squeezed into the seams of a wooden boat to make it watertight.

CENTERBOARD. A wooden or metal plate lowered into the water through a watertight housing in the bottom of the boat. It resists the boat's sliding sideways.

CHAFER. Replaceable material such as used netting or cowhide that is attached to the underside of a trawl to protect it from wear as it slides over the ocean's floor or rubs against the boat's side.

COD END. The end of a trawl net. The fish are eventually pushed into the cod end as the net is dragged along.

DAVIT. A fixed or movable crane that projects over the side of a boat or over a hatchway. It is used for hauling nets, anchors, boats or cargo.

DRAGNET. Another name for a trawl. Trawling vessels are often called dragboats.

DREDGE. Fishing gear made of a rectangular iron frame supporting a bag of fiber or wire webbing. The dredge is dragged along the bottom and is a common way of landing oysters and other mollusks.

DRY ROT. Wood decay caused by a type of fungus that grows on moist, unventilated areas.

FAIR-LEAD. A ring or piece of wood with holes in it that acts as a guide for a rope or cable. It is often used to guide the cable attached to an otter trawl.

FATHOM. Six feet.

FISHERY. An area that supplies an abundant amount of fish for commercial purposes.

FLYING BRIDGE. A raised platform, usually on top of the deckhouse, that affords good visibility.

FREEBOARD. Distance from the waterline to the railing.

GAFF. A pole with a large hook at its end.

GANGIONS (pronounced "gan-yuns"). The short branch lines attached to the main line of a setline. At the end of each gangion is a baited hook.

GILLNET. A net, usually rectangular, with the mesh size such that fish of the proper size that strike the net become stuck in the webbing. Very small mesh is used for fish such as herring, and larger mesh is used for salmon.

GROUNDFISH. Fish that spend most of their lives on or near the ocean's floor. Also called bottomfish.

GUNWALE (pronounced "gun-nel"). A boat's rail.

GURDIES. Spools used in trolling upon which the fishing line is wound. The gurdies are usually powered, but on some of the smaller boats, like salmon dories, they are often hand operated.

KEEL. A boat's backbone. It runs from stem to stern along the bottom of the boat.

OTTER TRAWL. A cone-shaped net that is dragged along the sea bottom. Its mouth is kept open by floats, weights and by two otter boards which shear outward as the net is towed.

PURSE SEINE. A net that is cast in a circle around a school of fish. When the fish are surrounded, the bottom of the net is closed up, preventing escape.

SCHOONER. A sailboat that usually has two masts, although some have up to seven. The largest of the two masts (the mainmast) is closer to the stern.

SCUPPER. A hole or opening in a rail that allows water to drain off and trash fish and entrails to be swept overboard.

SETLINE. Fishing gear made up of a long main line attached to which are a large number of short branch lines. At the end of each branch line is a baited hook. When catching groundfish, setlines are laid on the seafloor. When catching swordfish, shark or tuna they are buoyed near the surface. Setlines can be twenty or more miles long. They are also called longlines.

SHRIMP TRAWL. A small otter or beam trawl.

SLOOP. A single-masted sailboat whose mast is located more than one-third of the way from the bow to the stern.

TRIMMING. Adjusting a boat's load so that it rides in a certain way. Tuna clippers are trimmed so that the stern rides very low in the water, allowing fish to be easily hauled aboard.

TROLLING. Fishing with the use of lured or baited lines that are dragged through the water behind the boat.

TUBS. Wicker baskets that hold sections of setlines. The hooks of the setline are stuck into the basket's cork or soft rope rim.

Bibliography

Albin, Alexander and Alexander, Ronelle. *The Speech of Yugoslav Immigrants in San Pedro, California.* Publications of the Research Group for European Migration Problems, no. XVII, Netherlands, 1972.

Andrews, R. W. and Larssen, A. K. *Fish and Ships.* Superior Publishing Co., 1959.

Avery, M. *History and Government of State of Washington.* University of Washington Press, 1961.

Bailey, Paul. *City in the Sun.* Westernlore Press, 1971.

Bergmann, Leola Nelson. *Americans from Norway.* J.B. Lipincott Co., 1950.

Bjork, Kenneth O. *West of the Great Divide.* Norwegian-American Historical Association, 1958.

Blegen, Theodore C. *Norwegian Migration to America.* Norwegian-American Historical Association, 1940.

Bosworth, Allan R. *America's Concentration Camps.* W.W. Norton and Co., 1967.

Browning, Robert J. *Fisheries of the North Pacific.* Alaska Northwest Publishing Co., 1974.

Connolly, James B. *The Port of Gloucester.* Doubleday, Doran and Co., Inc., 1940.

Crandall, Julie V. *The Story of Pacific Salmon.* Binfords and Mort, 1946.

Crutchfield, J.A., and Pontecorvo, G. *The Pacific Salmon Fisheries.* The Johns Hopkins Press, 1969.

Digges, Jeremiah. *In Great Waters.* Macmillan Co., 1941.

Doering, J. Frederick. "Folk Customs and Beliefs of Greek Sponge-Fishers of Florida." *Southern Folklore Quarterly,* vol. 7, no. 2 (June 1943).

Doliber, Earl. *Lobstering Inshore and Offshore.* International Marine Publishing Co., 1973.

Finn, William. *Fishermen on Georges Bank.* Little, Brown and Co., 1972.

Gann, Ernest K. *Song of the Sirens.* Simon and Schuster, 1968.

Garner, John. *Modern Deep-Sea Trawling Gear.* Coward and Gerrish Ltd., England, 1967.

Georges, Robert A. "The Greeks of Tarpon Springs: An American Folk Group." *Southern Folklore Quarterly,* vol. 29, no. 2 (June 1965).

Jensen, Albert C. *The Cod.* Thomas Y. Crowell Co., 1972.

Laevastu, Taivo and Hela, Ilmo. *Fisheries Oceanography.* Coward and Gerrish, Ltd., England, 1970.

Lippa, E.J.R. "British Columbia Trawlers and Trawl Gear." Fisheries Research Board of Canada, Technical Report no. 13, 1975.

Marden, Luis. "The Sailing Oystermen of Chesapeake Bay." *National Geographic,* December 1967.

McKervill, Hugh M. *The Salmon People.* Evergreen Press, Ltd., Vancouver, British Columbia, 1967.

Mead, John T. *Marine Refrigeration and Fish Preservation.* Business News Publishing Co., 1973.

Merritt, J.H. *Refrigeration on Fishing Vessels.* Fishing News (Books) Ltd., London, 1969.

Netboy, Anthony. *The Atlantic Salmon.* Faber and Faber Ltd., London, 1968.

Rygg, A.N. *Norwegians in New York, 1825-1925.* Norwegian News Co., 1941.

Speroni, Charles. "California Fishermen's Festivals." *California Folklore Quarterly,* vol. 14, no. 2 (April 1955).

Traung, Jan-Olof. *Fishing Boats of the World,* vols. 1 and 2. Fishing News Ltd., England, 1960.

Vlassis, George D. *The Greeks in Canada.* Canada, 1953.

Warner, William W. "Winter 'Drudging' Lifts Crabs from Chesapeake Mud." *Smithsonian,* Feb. 1976.